I0748605

Foreword — A Singular Mission

I have one goal.

Not to scare parents.

Not to shame technology.

Not to pretend there's a single rule that fixes everything.

My goal is to get better information into the hands of families before something goes wrong.

Imagine a parent stepping into a situation already in motion. Something feels off. A message was deleted. A tone has shifted. A child is overwhelmed, and the adults are scrambling to catch up. Everyone is asking some version of the same question: *How did this get so complicated so fast?*

The answer is rarely mysterious.

No one ever explained the system.

Most parents were given tools without a map. They were told to turn on parental controls. To trust their gut. To have "the talk." What they weren't given was an explanation of how attention is shaped, how identity forms online, how incentives drive platform behavior, or how a dozen small, reasonable decisions can quietly stack into something risky.

They weren't given language for what they were noticing.

They weren't shown how to respond without overreacting.

They weren't taught how to hold boundaries without breaking trust.

So, they improvised. Most families do. And improvising under pressure is exhausting and dangerous.

This playbook exists because families deserve better than improvisation.

I believe most digital harm is preventable, not through perfect rules or constant monitoring, but through preparation. When parents understand what they're dealing with, they make different choices. They notice earlier. They stay calmer. They keep conversations open instead of shutting them down. And that changes outcomes in quiet ways that never make headlines, but matter deeply to the kids living them.

This isn't about raising tech experts.

It's about raising children who can think clearly inside systems designed to make that harder.

It's about helping parents move from reaction to intention. From fear to understanding. From rules that spark conflict to reasoning that builds trust.

If this work helps one family pause before panic, ask a better question, or handle a hard moment without losing connection, then it's doing what it's meant to do. If it helps many families do that, the mission simply grows.

Everything here serves that purpose: giving parents a clear, usable understanding of the digital world their children are growing up in, so fewer families have to learn the hard way.

That's the work.

That's the point.

That's why this exists.

Most families aren't careless. They're uninformed. And that's fixable.

Matthew McBride

Preface — Why This Playbook Exists

I didn't start thinking about kids' digital lives because I wanted to police screens.

I started because I kept seeing the same pattern: families doing their best, using the tools they were told would "keep kids safe," and still ending up confused, exhausted, and blindsided when something went sideways. Not because they were careless. Because the system they're parenting inside of is built to be difficult.

I work in digital forensics. That means I live on the wrong end of "it'll probably be fine." I see what happens when privacy settings don't get touched, when a chat app becomes a private hallway, when a simple "send me a pic" turns into pressure, and when "we'll deal with it later" turns into "we can't undo that." I also see the opposite: families who respond steadily, preserve what matters, don't panic, and get their kid through something hard without turning the house into a courtroom.

This playbook exists because most guidance about digital safety falls into one of two useless extremes.

One extreme is fear. Everything is a threat. Every app is a predator portal. Every kid is one click away from ruin. That approach doesn't protect families—it burns them out. It trains kids to hide, not to ask for help.

The other extreme is denial dressed up as optimism. "Kids will figure it out." "They're digital natives." "Just talk to them." Talking matters, but talking alone can't compete with design teams whose entire job is to keep your child's attention, shape their habits, and turn their feelings into revenue.

Families need something better than panic or vibes.

They need science *and* compassion.

Science, because the modern internet isn't neutral. It's engineered. It runs on incentives: engagement, retention, growth. Your child is not weak for struggling with a system designed to be sticky. Your teen is not broken because they can't stop scrolling when the interface removes every natural stopping point. If you understand the incentives

and the mechanics, a lot of "mystery behavior" stops being mysterious.
Compassion, because kids aren't just users of technology—they're developing humans. Their impulse control is still under construction. Their social world is high-stakes. Their need to belong is not a bug; it's a feature of being young. If your approach to safety humiliates them, interrogates them, or makes honesty feel dangerous, you'll lose the one thing you can't replace with any tool: access to the truth.

Understanding the system doesn't make parenting easy. It makes it possible.

This playbook is designed to help you reclaim control without becoming controlling.
Not by banning everything.
Not by trusting everything.
Not by trying to outsmart the internet.
But by building a household system that holds up under real life: busy weeks, tired parents, stressed kids, and apps that never sleep.

You'll find three recurring ideas throughout this book—because some things are worth repeating until they stick:

First: Your goals and the platform's goals are not aligned.
Sometimes they overlap. Often they don't. Acting like they do is how families drift into defaults they didn't choose.

Second: Friction is not cruelty. Friction is safety.
A second channel check. Phones out of bedrooms. Notifications turned off. A weekly check-in. These aren't punishments; they're guardrails. In a world optimized for impulsive clicks, a little speed bump is mercy.
Third: "When, not if."
Something will slip through. Your child will see something. Someone will message them. A boundary will get tested. Your job isn't to build a perfect wall. It's to build a relationship and a response plan that makes problems survivable and learnable.

If you're reading this because you're worried—because something already happened, or because you can feel how fast the landscape is changing—take a breath.
You don't need to become a cybersecurity professional to parent well in a digital world. You just need a clearer model of what you're dealing with, a few reliable practices, and a steady posture that your kids can trust.

This playbook is a map. Not for a perfect family. For a real one.
Let's build something durable.

Chapter 1

The Attention Economy:
Why Screens Feel Impossible To Put Down

"If attention is constantly pulled outward, focus doesn't disappear, it never gets a chance to form."

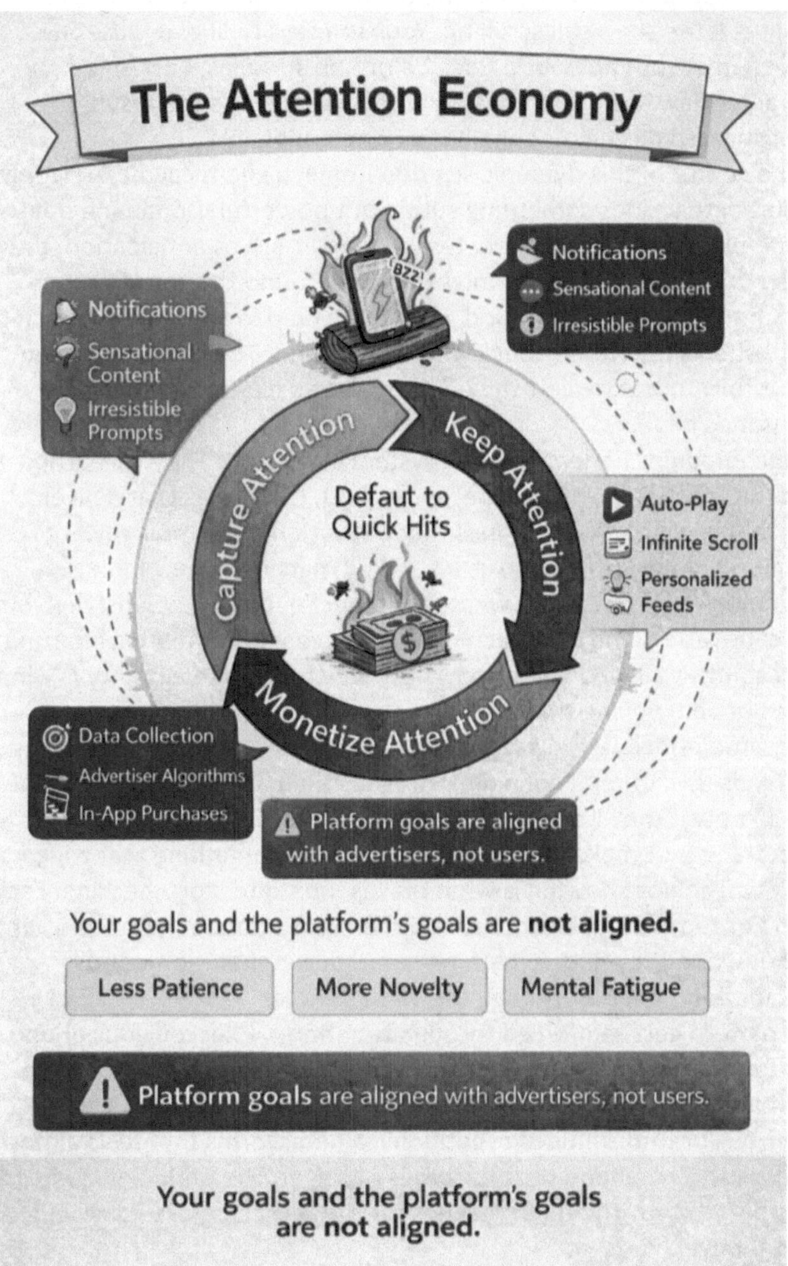
The Attention Economy
Notifications
Sensational Content
Irresistible Prompts
BZZ
Notifications
Sensational Content
Irresistible Prompts
Capture Attention
Keep Attention
Defaut to
Quick Hits
Auto-Play
Infinite Scroll
Personalized
Feeds
Monetize Attention
Data Collection
Advertiser Algorithms
In-App Purchases
Platform goals are aligned
with advertisers, not users.
Your goals and the platform's goals are **not aligned.**
Less Patience
More Novelty
Mental Fatigue
Platform goals are aligned with advertisers, not users.
Your goals and the platform's goals
are not aligned.

1.1 The Dopamine Loop

Your daughter picks up her phone to check one notification. Twenty minutes later, she's still scrolling. Your son says he'll play "just one more game," but an hour passes before he surfaces. This isn't a character flaw. It's not laziness or disobedience. It's the result of sophisticated design meeting basic human biology.

At the center of this dynamic sits dopamine, a chemical messenger in the brain that does something subtle but powerful. Dopamine doesn't create pleasure exactly. It creates wanting. It drives anticipation, the sense that something good might be just around the corner. When early humans foraged for food, dopamine helped them keep searching even when they weren't sure what they'd find. That system worked beautifully for survival. It still exists in all of us, unchanged for thousands of years.

Tech companies understand this system intimately. They've learned that the most compelling experiences aren't the ones that deliver constant rewards, but the ones that deliver *unpredictable* rewards. Imagine a slot machine versus a vending machine. The vending machine gives you exactly what you expect every time. Satisfying, but not captivating. The slot machine might give you something amazing, or nothing at all, and you never know which. That uncertainty is what keeps people pulling the lever.

This same principle shows up everywhere in the apps and platforms your kids use. When your child refreshes their social media feed, they don't know if they'll see something boring, something funny, something that makes them feel included, or something that triggers anxiety. The not knowing is what makes it hard to stop checking. Each swipe or tap is a small gamble, and the brain's dopamine system lights up not when they find something good, but in that moment of anticipation right before they see what's there.

The pattern gets reinforced through repetition. Checked your phone and found something interesting? Your brain notes that. Didn't find anything? The uncertainty keeps you curious for next time. Over hundreds or thousands of repetitions, a habit forms that feels almost automatic. The phone buzz becomes a trigger. The brain anticipates a possible reward. The hand reaches for the device before conscious thought even kicks in.

What makes this particularly challenging for kids is that their brains are still developing the circuitry that helps them pause and think through whether they actually want to do something. The part of the brain that experiences wanting matures earlier than the part that helps with self-regulation and long-term thinking. So children and teenagers feel the pull of these dopamine-driven loops intensely, but they're working with less-developed tools for resisting them.

None of this means screens are evil or that dopamine is bad. Dopamine helps us learn, motivates us to try new things, and makes life feel engaging rather than flat. The issue isn't the chemical or even the technology in isolation. The issue is the mismatch between design systems optimized to capture as much attention as possible and developing brains that are still learning how to manage their own impulses.

What rarely gets discussed is recovery. Dopamine systems don't just get overstimulated; they need quiet to recalibrate. Boredom isn't a failure state. It's the brain catching its breath

When you understand this, the seemingly inexplicable behaviors start to make sense. Your child isn't choosing the screen over family dinner because they don't care about you. They're experiencing a genuinely powerful biological pull that's been carefully cultivated through design. Recognizing this doesn't excuse everything, but it does change how we respond. Instead of seeing defiance or disrespect, we can see someone learning to navigate forces that even adults find difficult to resist.

The good news is that understanding the dopamine loop gives us insight into what might actually help. We're not trying to eliminate dopamine or remove all pleasure from technology. We're trying to help kids recognize when they're being pulled by that anticipation cycle, and to build the capacity to notice it happening and sometimes choose differently.

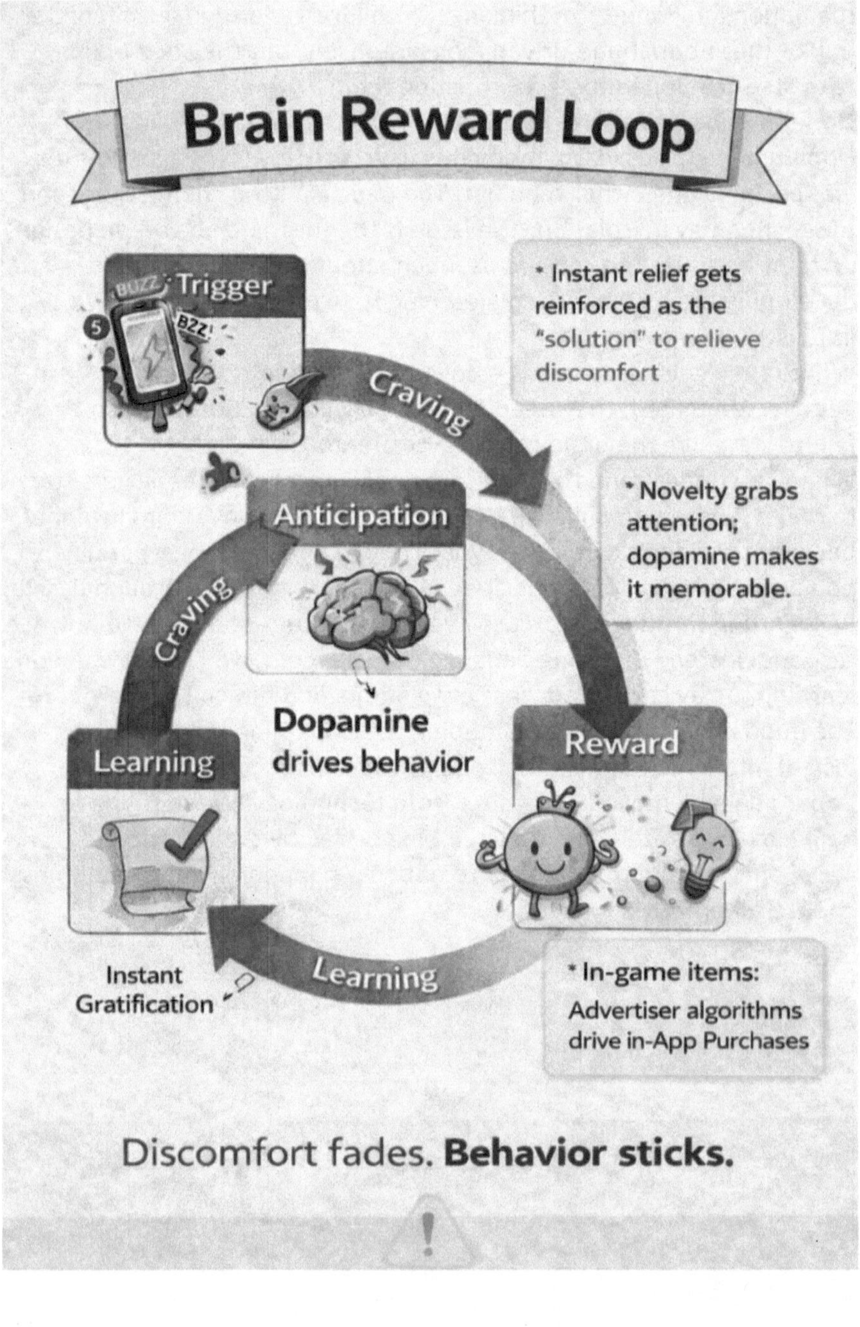

Brain Reward Loop

BUZZ Trigger

5

BZZ!

Craving

* Instant relief gets reinforced as the "solution" to relieve discomfort

Anticipation

* Novelty grabs attention; dopamine makes it memorable.

Craving

Dopamine
drives behavior

Learning

Reward

Instant Gratification

Learning

* In-game items:
Advertiser algorithms drive in-App Purchases

Discomfort fades. **Behavior sticks.**

1.2 Variable Reward Systems

Your daughter checks Instagram. Nothing new. She refreshes. Still nothing. She checks again thirty seconds later—and there it is: three new likes, a comment from her best friend, and a follow request from someone at school. Her brain lights up with a little burst of satisfaction.

This pattern repeats dozens of times a day, and it's not accidental. The unpredictability is the point.

Back in the 1930s, a psychologist named B.F. Skinner put pigeons in boxes with levers. When a pigeon pressed the lever, sometimes it got food. Sometimes it didn't. Skinner discovered something unexpected: the pigeons pressed the lever far more obsessively when the reward was unpredictable than when it came every single time. The uncertainty itself became irresistible. The pigeons would keep pressing, keep checking, driven by the possibility that this time might be the time.

Tech companies didn't invent this mechanism, but they've become extraordinarily good at deploying it. Every time your child pulls down to refresh a feed, they're essentially pressing that lever. Maybe there will be a new message. Maybe someone liked their post. Maybe something interesting appeared in the last three minutes. The maybe is what makes it so hard to stop.

This is what's meant by a variable reward system. The reward—a notification, a like, a new video, a message—arrives on an unpredictable schedule. Sometimes you get something immediately. Sometimes you wait. Sometimes you get more than you expected. Your brain never quite knows what's coming next, so it stays in a state of anticipation. That anticipation feels a lot like excitement, even when the actual content is mundane.

Think about the difference between checking your mailbox and checking your phone. Your mail arrives once a day, at roughly the same time. You don't feel compelled to check it every ten minutes because you know nothing will have changed. But your phone? Your phone might have something new right now. Or now. Or now. The uncertainty keeps you coming back.

Social media platforms build their entire notification systems around this principle. They don't send you alerts the moment something happens. They batch them, delay them, and release them in ways designed to bring you back when engagement is dropping. You might get one notification, then two more a few minutes later, then nothing for an hour, then a sudden cluster. The pattern feels random, but it's carefully managed to keep you guessing and checking.

The same logic shapes the feeds themselves. Algorithmic timelines don't show you posts in chronological order anymore. They shuffle content, prioritize some posts over others, and insert new items at unpredictable intervals. You keep scrolling because you never quite know what's coming next. Maybe it'll be something from a close friend. Maybe it'll be something funny. Maybe it'll be something you didn't even realize you wanted to see. Every swipe is another pull of the lever.

This isn't about your child lacking willpower or discipline. Variable reward systems work on everyone. They work on adults who know exactly how they function. They work on researchers who study them. The pull is neurological, not moral. When dopamine pathways are engaged by unpredictability, the urge to check becomes genuinely difficult to resist.

What makes this especially challenging for families is that these systems are invisible. Your child isn't thinking, "I'm being manipulated by a variable reward schedule." They're thinking, "I just want to see if anyone texted me back." The compulsion feels like it's coming from within, like a personal desire or a social need. And in some ways it is—but it's also being amplified and exploited by design choices that have nothing to do with your child's actual relationships or interests.

You'll notice this in how your child uses their phone. They check it in the gaps: between bites of dinner, during a pause in conversation, in the three seconds before a video loads. They're not necessarily looking for anything specific. They're checking because the possibility exists. The habit becomes automatic, wired into the small spaces of daily life.

Understanding this doesn't mean you need to eliminate all apps or ban all notifications. But it does help explain why "just don't look at it so much" doesn't work as advice. The design is meant to override that kind of intention. Willpower runs out. Systems don't.

What you're up against isn't a lack of self-control. It's a deliberately engineered environment that understands human psychology better than most of us understand it ourselves. Recognizing that won't solve everything, but it shifts the conversation. The question isn't why your child can't put the phone down. The question is how to build practices and boundaries that acknowledge the pull and work with it, rather than pretending it doesn't exist.

1.3 Micro-Validation and Social Currency

When your daughter posts a photo and checks back three minutes later to see who's liked it, she's not being vain or shallow. She's responding to something that runs much deeper than a number on a screen.

Every like, heart, or flame is a tiny social signal. In the physical world, we've always relied on these signals to understand where we stand with other people. A smile across the cafeteria. An invitation to sit at a lunch table. A friend who saves you a seat. These moments tell us we're seen, we're valued, we belong. Online platforms have taken this fundamental human need and made it quantifiable, instant, and visible to everyone.

The difference is scale and speed. In earlier generations, you might wonder whether people enjoyed your performance in the school play or liked your new haircut, but you'd find out gradually, through scattered conversations and subtle cues. Today, the feedback arrives in real time, counted and displayed. It's no longer ambiguous. Sixty-three people liked your post. Twelve people didn't respond to your story. Your streak with your best friend just hit 487 days.

This creates what researchers call micro-validation: small, frequent hits of social approval that feel good in the moment but don't necessarily build lasting confidence. Think of it like eating handfuls of cereal straight from the box instead of sitting down to a meal. Each handful gives you something, but you're never quite satisfied, and you keep reaching for more.

The mechanism behind this involves dopamine, a neurotransmitter that helps regulate motivation and reward. When something uncertain might result in a positive outcome, dopamine levels rise. That rising feeling is what keeps us checking, refreshing, waiting to see what happens next. Slot machines work on the same principle. You don't know if the next pull will pay out, so your brain stays engaged, alert, hoping. Social media notifications operate on a similar variable reward schedule. Sometimes your post gets a lot of attention. Sometimes it doesn't. You can't predict it, so you keep checking.

For kids and teens, whose brains are still developing and whose sense of identity is actively forming, this system hits especially hard. They're already navigating complicated questions about who they are and whether they matter. They're watching how they're perceived, testing out different versions of themselves, trying to figure out where they fit. When the feedback loop is public, instant, and numerical, it can start to feel like the metrics are the answer. If this post got more likes than the last one, maybe I'm doing something right. If no one responded to my story, maybe I said the wrong thing.

Over time, this can shift how young people think about their own worth. Instead of developing an internal sense of self based on their values, interests, and relationships, they begin outsourcing that evaluation to an audience. They start to curate not just what they share, but who they are, based on what seems to generate positive responses. A girl who loves painting might stop posting her artwork because it doesn't get as much attention as photos of her at parties. A boy might adopt a persona that feels exaggerated or performative because it gets more engagement than his real thoughts.

The stakes feel higher because the audience is always there. In past decades, you could have a bad day at school, come home, and reset. The social world paused. Now it continues around the clock. Your phone buzzes with reactions and messages at dinner, during homework, in bed. If you don't respond quickly enough to a friend's message, you might worry you've hurt them or missed something important. If you don't maintain your streaks, you risk losing a visible symbol of closeness. The pressure to stay connected and responsive becomes constant.

Parents often notice their child's mood shifting in sync with their phone activity. A teenager who seemed happy suddenly looks deflated after scrolling. A middle schooler gets anxious if they can't check their messages for an hour. These aren't signs of weakness or poor character. They're natural responses to a system designed to keep users engaged by tapping into core emotional needs.

What makes this particularly tricky is that social connection is genuinely important. Humans are wired to care what others think, to seek belonging, to feel recognized. The instinct itself is healthy. The problem arises when the platforms that facilitate connection also gamify it, turning relationships into metrics and presence into performance. Kids can end up feeling like they're constantly auditioning for approval, even from people they already know and trust.

The concept of social currency helps explain why this feels so compelling. Currency implies value, trade, status. Online, your likes and followers and engagement become a form of capital. They signal to others that you're worth paying attention to. They can open doors to new friendships, romantic interest, social opportunities. Losing that currency, or seeing someone else accumulate more of it, can feel like a real loss, even if nothing tangible has changed.

This doesn't mean kids are doomed or that platforms are purely manipulative. Many young people maintain healthy skepticism about online validation and build meaningful relationships through digital tools. But it does mean that the environment they're growing up in is different from the one most parents experienced, and it requires a different kind of awareness.

When you understand that your child's attachment to their phone isn't about defiance or distraction, but about a deeply human need to feel connected and valued, it changes how you might respond. The goal isn't to dismiss those needs or shame them for caring what others think. It's to help them build a more stable foundation for their sense of self, one that doesn't rise and fall with each notification. That work starts with recognizing what's actually happening beneath the surface.

1.4 Overstimulation and Focus Erosion

Imagine a child who used to play contentedly with blocks for twenty minutes at a stretch. Now, after six months of daily tablet time, those same blocks hold her attention for maybe three minutes before she's asking what's next. Her parents haven't changed their parenting. The blocks haven't gotten less interesting. But something fundamental has shifted in how her brain approaches the world.

What's shifted is her baseline expectation for stimulation.

Digital platforms are designed to deliver what researchers call "supernormal stimuli." These are experiences more intense, more colorful, more rewarding than anything humans evolved to encounter naturally. A toddler app delivers a burst of confetti and a cheerful sound every few seconds. A video for school-age kids cuts to a new image every two to three seconds. A social feed offers an endless scroll of novel content, each post engineered to trigger just enough interest to keep you moving to the next one.

The human brain is built to adapt to its environment. When a child spends hours each day in a world of rapid cuts, instant rewards, and constant novelty, their nervous system recalibrates. It begins to expect that pace as normal. Anything slower starts to register as boring, not because it lacks value but because the brain has grown accustomed to a level of stimulation the real world simply doesn't provide.

You can observe this recalibration in surprisingly concrete ways. Picture a child finishing thirty minutes on a device. When it's time to transition to something analog, drawing or playing outside, they often seem unable to settle. They're restless, irritable, complaining of boredom in a way that feels almost physical. What looks like a behavioral problem is actually a neurological one. Their system has been running at a heightened pitch, flooded with dopamine and sensory input. Now that the stimulation has stopped, there's a crash. The brain doesn't have a neutral gear ready anymore.

Over time, this pattern doesn't just create friction around screen transitions. It begins to reshape how children approach everything that requires patience. Consider the child who used to work through a frustrating puzzle or sit with a picture book, tolerating the slow build of a story. After months of high-intensity digital input, those same activities can feel unbearable. The child isn't being defiant. Their brain has simply practiced a different kind of attention, one that expects constant reward and rapid change.

This matters beyond childhood tantrums or complaints about boredom. Sustained attention is the foundation for almost all meaningful learning. When you're figuring out how fractions work or why a character in a story made a particular choice, you need to sit with confusion. You need to tolerate the discomfort of not understanding yet. But if your brain has been trained to expect immediate feedback and frequent novelty, that tolerance window shrinks. The moment something feels hard or slow, the impulse is to swipe away, to find something easier, to avoid the discomfort entirely.

Adolescents who grow up in high-stimulation environments often describe feeling chronically underwhelmed by real life. A conversation with a friend can feel painfully slow. A family dinner drags. Hobbies that used to absorb them now seem dull. The world hasn't become less interesting, but their brains have become dependent on a level of input that everyday life doesn't naturally offer. That gap leaves them reaching for their phones not because there's something specific they want to see, but because the absence of stimulation itself feels intolerable.

There's also a quieter cost that's harder to measure. Deep focus, the kind that lets you lose yourself in a project or a book, requires surrendering vigilance. You have to stop scanning for the next thing. But when a brain has spent years monitoring for notifications, for new posts, for the next reward, that surrender becomes much harder. Even when the phone isn't present, the habit of fractured attention lingers. Kids describe it as feeling like their thoughts are slippery, like they can't hold onto an idea long enough to think it through. They're not imagining it. The neural pathways for sustained attention have been under-practiced while other pathways, the ones for rapid task-switching and novelty-seeking, have been reinforced thousands of times over.

None of this means screens have permanently damaged a generation. Brains remain adaptable throughout childhood and adolescence. Attention isn't destroyed, it's out of shape. With consistent time in lower-stimulation environments, kids can rebuild their capacity for sustained focus. But it doesn't happen quickly, and it doesn't happen automatically. It requires creating space where boredom is allowed to last long enough for a child's own curiosity to come back online.

That process can be uncomfortable for everyone involved. A child whose brain expects constant input will genuinely struggle when that input stops. They're not being manipulative when they say they're bored. They're describing a real neurological state, one where their system hasn't yet learned to generate its own momentum. Parents often interpret this discomfort as a sign they need to provide more activities, more enrichment, more entertainment. But sometimes the most important thing you can provide is less. Less input, less stimulation, less rescue from the feeling of boredom itself.

What we're really talking about is the ability to be present in your own life. To sit through a conversation without reaching for a distraction. To stay with a problem long enough to solve it. To tolerate the slow, iterative, sometimes tedious process of getting good at something. These aren't old-fashioned values. They're the cognitive foundations that allow people to learn, to connect, and to pursue anything worth pursuing. And they develop through practice in environments that don't do the regulating for you.

A tired brain doesn’t need more discipline. It needs fewer inputs.

1.5 Cognitive Load and Attention Recovery

There's a moment many parents recognize: your child finishes a show, closes a game, or puts down their phone, and for the next stretch of time they seem unable to land anywhere. They wander from room to room. They say they're bored, but they reject every suggestion you offer. They seem restless and irritable, as if something is missing but they can't say what.

What's happening in that moment isn't defiance or ingratitude. It's a kind of cognitive exhaustion that we're only beginning to understand. The brain has been running at high speed, processing rapid cuts, bright colors, notifications, and decision points. When that stimulation stops, it doesn't smoothly shift into a calmer mode. It stalls.

Think of it like stepping off a treadmill that's been moving fast. Your legs still feel like they're moving even though the ground is still. The brain works the same way. After sustained exposure to high-stimulation content, it takes time for the nervous system to recalibrate. Researchers call this cognitive load, and when it's chronic, it doesn't just make kids irritable in the moment. It changes how attention works over time.

Attention isn't a fixed trait. It's more like a muscle that gets trained by how we use it. When a child spends hours each day in environments designed to deliver novelty every few seconds, their brain begins to expect that pace. Anything slower starts to feel unbearable. A book seems too slow. A conversation feels effortful. Even play that used to be engaging now feels flat unless it includes a screen.

This isn't about intelligence or willpower. It's about what the brain learns to crave. High-stimulation media works by triggering small bursts of dopamine, the neurotransmitter associated with reward and motivation. Games, apps, and videos are engineered to deliver these hits frequently and unpredictably, which keeps the brain leaning forward, waiting for the next one. When that pattern repeats day after day, the brain's reward system recalibrates. It starts to need more stimulation to feel engaged, and less stimulating activities stop registering as worthwhile.

The other casualty is patience. Patience isn't passive waiting. It's an active cognitive skill that involves tolerating discomfort, delaying gratification, and staying present with something that hasn't yet delivered a payoff. A child building with blocks has to hold an idea in mind, adjust when something doesn't work, and keep going even

when progress is slow. That process strengthens the prefrontal cortex, the part of the brain responsible for planning, focus, and self-regulation. But when a child's day is filled with instant feedback loops, those neural pathways don't get the same workout.
The strange thing is that boredom used to do some of this work. Not the frustrated, overstimulated kind of boredom that leads to meltdowns, but the quieter version, the kind where a child is understimulated just enough that their brain starts looking for something to do. That's when creativity kicks in. A stick becomes a sword. A cardboard box becomes a spaceship. The child's internal world takes over because the external world isn't handing them a script.
But boredom has become rare. There's almost always a device within reach, and the gap between "I have nothing to do" and "I'm watching something" has collapsed to seconds. The brain never has to generate its own entertainment, solve its own problem of what to do next, or sit with the low-level discomfort of not being occupied. Over time, that skill atrophies.
This isn't irreversible. Attention can recover, but it takes time and it requires creating space where the brain isn't constantly fed input. Imagine a parent who starts building in short daily windows where screens aren't an option and no structured activity is planned. At first, the child might resist. They might complain, pace, or seem genuinely distressed. But if the parent holds the boundary calmly and without judgment, something begins to shift. The child starts noticing things. They pick up a book they've ignored for months. They start a project. They get interested in something small and odd, the kind of thing that only appears when there's nothing louder competing for attention.
This process doesn't happen overnight, and it doesn't look the same for every child. Some kids take days to adjust, others take weeks. But the pattern is consistent: when the environment changes, the brain adapts. The key is that the parent has to believe it's worth the discomfort of the transition period, and the child has to experience enough repetition that their nervous system learns to expect and tolerate lower levels of stimulation.
One other thing that helps is modeling. If a parent is constantly on their phone during downtime, the child's brain registers that as the norm. But if the parent is reading, cooking, tinkering with something, or just sitting quietly, the child's brain gets a different message: that it's

possible to be calm and engaged without a screen. Kids don't need lectures about attention. They need to see what restored attention looks like in the people around them.

What we're really talking about is giving the brain a chance to slow down enough that it can notice what it actually wants, not just what it's been trained to reach for. When that happens, focus isn't something a child has to force. It returns on its own, quietly, the way sleep comes when the room finally gets dark.

1.6 The Pocket Slot Machine

Imagine walking through a casino at two in the morning. The lights pulse in rhythmic patterns. Someone sits hunched at a machine, pressing a button, watching the reels spin, waiting to see what comes up. Press, spin, check. Press, spin, check. They're not necessarily having fun. They might not even want to be there anymore. But something about that next pull keeps them going.

Now picture a teenager on the couch, thumb scrolling through a feed. Swipe, load, glance. Swipe, load, glance. They're not looking for anything in particular. If you asked what they saw thirty seconds ago, they might not remember. But the hand keeps moving.

The resemblance isn't coincidental. Many of the engineers who design social media feeds, notification systems, and recommendation algorithms studied the same behavioral psychology that shaped gambling machines. They learned what keeps people coming back, what makes it hard to stop, and how to deliver rewards on a schedule that feels random enough to stay interesting.

This isn't about blaming your child or shaming them for lacking willpower. It's about recognizing that some digital experiences are engineered specifically to be hard to put down, using techniques refined over decades to hold human attention.

The core mechanism is something called variable ratio reinforcement. It sounds technical, but the idea is straightforward. If you got something good every single time you checked your phone, the behavior would feel predictable and lose its pull. If you never got anything good, you'd stop checking. But when rewards arrive unpredictably—sometimes right away, sometimes after several tries, sometimes not at all—the brain stays engaged in a way that's difficult to resist. You keep checking because this time might be the time.

Think about the difference between a vending machine and a slot machine. You put money in a vending machine, press a button, and your snack drops. Predictable, functional, done. But a slot machine works differently. Sometimes you win a little. Sometimes nothing. Sometimes a near miss that almost paid off. The uncertainty is the engine. And your phone, when it's delivering social feedback or surfacing content through an algorithmic feed, often works more like the slot machine than the vending machine.

Every time a notification arrives, there's a micro-moment of curiosity. Who is it? What do they want? Is it interesting? The answer varies. Sometimes it's a close friend. Sometimes it's a group chat gone off the rails. Sometimes it's nothing that matters. That unpredictability keeps attention tethered to the device in a way that a predictable, useful tool would not.

The same pattern shows up in autoplaying videos, infinite scroll, and the way "likes" or comments trickle in at irregular intervals. The design isn't neutral. It's shaped to maximize something the industry calls engagement, which in practical terms means time spent and interactions logged. The business model depends on it.

None of this means screens are evil or that technology is inherently corrupting. It means that some features are designed with a specific goal, and that goal isn't always aligned with how you might want your child to spend their time or attention. A calculator app doesn't work this way. A word processor doesn't work this way. But a social feed often does.

Parents sometimes feel confused when a child says they want to stop scrolling but can't seem to put the phone down. It looks like a lack of self-control, maybe even defiance. But if you understand the design, it makes more sense. The child isn't choosing the screen over family dinner because they don't care. They're caught in a feedback loop that was built to be difficult to exit. That doesn't erase responsibility, but it does change how you might think about the situation.

It also helps explain why some of the most common advice doesn't work as well as it should. Telling a kid to "just use it less" is like telling someone at a slot machine to "just pull the lever less." Technically possible, sure. But it misses the point. Willpower is real, but it's not unlimited, and it's especially hard to summon when you're up against a system designed to deplete it.

Understanding this doesn't mean surrendering to the technology or deciding nothing can be done. It means recognizing what you're dealing with. If you know the machine is built to hold attention, you can start thinking about how to change the context around it—where the phone lives, when it's accessible, what replaces it when it's not there. You can talk with your child about how the design works, which sometimes makes the invisible visible and gives them a little more room to make intentional choices.

The goal isn't to demonize phones or pretend we can go back to a pre-digital world. It's to see the tools clearly, understand how they're shaped, and make decisions based on that understanding rather than vague frustration or guilt. The slot machine metaphor isn't perfect, but it gets at something true: some of what looks like a personal failing is actually a design working exactly as intended.

Chapter 2

Emotional Hijacking:
How Apps Shape Feelings And Identity

"Identity forms fastest where feedback is loudest."

2.1 The Online Mirror Effect

When your daughter posts a photo and checks back ten minutes later to see how many likes it got, she's not just being vain or obsessed. She's doing something deeply human: trying to figure out who she is by watching how others respond to her.

This is how identity has always formed. Babies learn to smile because adults smile back. Toddlers test boundaries and watch faces to see what's acceptable. Teenagers try on different versions of themselves and gauge reactions from friends, teachers, crushes. We become ourselves, in part, by seeing ourselves reflected in other people's eyes.

The difference now is the mirror.

It used to be that a kid might share a drawing with a parent, tell a joke to three friends at lunch, or wear a new shirt to school and notice who commented. The feedback was limited, delayed, and came from people they actually knew. If the reaction was lukewarm, there was time to shrug it off. If it was harsh, a friend might pull them aside later and soften the blow.

Online, the mirror is immediate, quantified, and everywhere. A twelve-year-old posts a video and within seconds can see a number. That number feels like a verdict. Worse, the number keeps updating. It ticks up or it doesn't. It might plateau at seven while a friend's similar video climbs to two hundred. The comparison is automatic and impossible to ignore.

What makes this especially powerful is that the feedback doesn't just come from peers anymore. It comes from an algorithm trained to maximize engagement. Imagine a teenager posting a funny video that gets modest attention. Then they post something edgier, angrier, or more vulnerable, and suddenly the algorithm picks it up. The views explode. The comments flood in. What they learn in that moment isn't just "people like this." They learn "this version of me is the one that matters."

The algorithm doesn't care if that version is healthy. It doesn't care if it's true. It only cares if it keeps people watching.

Over time, kids begin to optimize for the mirror. They don't just share what they think or feel. They share what they believe will perform. They start to notice patterns: vulnerability gets comments, outrage gets shares, certain aesthetics get follows. They adjust. They test. They iterate. This isn't cynical or manipulative. It's adaptive. They're trying to be seen, to matter, to figure out who they are in a system that rewards some signals and ignores others.

The trouble is that this process can warp the very thing it's supposed to reveal. Identity formation requires experimentation, mistakes, and the freedom to be inconsistent. A kid needs to try on different interests, change their mind, be awkward, fail privately. But if every attempt is immediately reflected back as a score, and if the scores start to define their social standing, the stakes get too high. They stop experimenting and start performing.

Consider a situation where a quiet kid discovers they're funny online. The comments pour in. For the first time, they feel confident, visible, maybe even popular. That's a good thing, up to a point. But if the only place they feel that way is in the comment section, they might start to believe the online version is the real one. The kid at the dinner table, the one their parents see, starts to feel like a rough draft.

Or imagine a teenager who posts about a hard day and gets dozens of supportive replies. That support is real, and it can be deeply comforting. But if they start to notice that posts about struggling get more engagement than posts about doing okay, they might begin to lean into the struggle. Not because they're faking it, but because the mirror has taught them that pain is the part of them that people care about most.

None of this means kids are passive victims. They're incredibly savvy. Many of them understand, on some level, that the algorithm is a game. But understanding it intellectually doesn't make them immune to it emotionally. Even adults who know better still feel a little spark when a post does well or a little sting when it doesn't. For a young person still figuring out who they are, those sparks and stings aren't just feedback. They're data points in an ongoing calculation about self-worth.

The online mirror also reflects a curated version of everyone else. Your son sees his classmates looking effortlessly happy, confident, successful. He doesn't see the twenty takes it took to get that photo or the caption that was rewritten six times. He just sees the final version and compares it to his own unfiltered life. The mirror tells him he's falling behind.

What makes this particularly hard for parents is that you can't see the mirror the way your child does. You see the posts, maybe, but you don't see the invisible scoreboard. You don't see them refreshing to check if that number went up. You don't see the sinking feeling when a friend's post outperforms theirs. You don't see them deleting something that didn't land, wondering what's wrong with them.

The question isn't whether kids should be online. For most families, that ship has sailed. The question is how to help them develop a sense of self that isn't entirely dependent on the mirror. How to make sure they know that the numbers don't tell the whole story. That the algorithm's preferences aren't the same as the truth. That the person they are when no one is watching, when nothing is being measured, is the person who matters most.

That's harder to teach when the mirror is always on.

2.2 The Comparison Trap

Your daughter scrolls through her feed and sees a friend's post from the beach: perfect lighting, everyone laughing, the kind of moment that looks effortless. What she doesn't see is the twenty takes it took to get that shot, or the argument that happened ten minutes later, or the fact that the friend felt anxious the whole afternoon. She just sees the final image. And something inside her whispers: *everyone else is having a better time than me*.

This is the comparison trap, and it operates with remarkable efficiency. Social platforms aren't designed to show us the full texture of other people's lives. They're designed to show us highlights, the peaks, the moments worth posting. We've always compared ourselves to others—that's human—but we used to do it with more complete information. You could see your neighbor's nice car, but you also saw them arguing on the front lawn or struggling to carry in groceries after a long day. The picture was more whole.

Online, the picture is never whole. It's curated, filtered, and optimized. A teen sees classmates at parties they weren't invited to. They see vacation photos, award announcements, relationship milestones. They see people who seem funnier, more confident, more liked. What they don't see is the person behind the screen, lying in bed at night feeling lonely or inadequate or wondering if they're good enough. Everyone is performing, but the performance looks like reality.

The psychological term for this is "upward social comparison," the act of measuring yourself against people who seem to be doing better. It's a normal part of growing up, but social media turns the volume way up. Instead of comparing yourself to a handful of peers you see in person, you're comparing yourself to everyone in your school, everyone you used to know, everyone you follow, and all the strangers the algorithm decides to show you. The scale is inhuman.

For kids and teens, whose sense of identity is still forming, this creates a relentless pressure. They start to believe that everyone else has figured something out that they haven't. They feel like they're falling behind in a race they didn't know they'd entered. And because they're seeing a steady stream of other people's best moments, their own ordinary days start to feel like evidence of failure.

Consider a middle schooler who posts a photo and then watches the likes trickle in. She sees a classmate's similar post get three times as many. Suddenly, it's not just about the photo. It's about her worth. It's about whether people like her, whether she matters, whether she's as interesting or attractive or funny as someone else. The platform has turned a social interaction into a ranked competition, complete with a visible scoreboard.

Parents sometimes wonder why their child seems so focused on how they look or what they're wearing or who's doing what. Part of that is adolescence, but part of it is the environment they're navigating. When your social world is mediated by images, and when those images are constantly being judged and ranked, appearance and presentation become survival skills. Looking good isn't vanity. It's currency.

The trap gets tighter because kids often respond to feeling inadequate by curating their own lives even more carefully. They stage their own photos, craft their own captions, project their own version of effortless success. Which means they're now contributing to the same cycle that's making them feel bad. They become both the audience and the performer, trapped in a feedback loop where no one is being real, but everyone is trying to look like they are.

This isn't about blaming kids for caring too much or being too sensitive. The platforms are engineered to exploit our need for social connection and validation. They surface content that triggers comparison because that keeps people scrolling. They reward certain kinds of posts with more visibility, shaping what kids think is normal or desirable. The system is working exactly as intended. It's just that the intended outcome isn't psychological wellbeing.

Some parents try to counter this by telling their kids that social media isn't real, that everyone's just showing their best side. That's true, but it's also not enough. Knowing intellectually that something is curated doesn't stop it from affecting you emotionally. Your daughter might understand that her friend's life isn't as perfect as it looks online, but that doesn't make her feel less left out when she sees photos from a hangout she wasn't at. Logic and feeling operate on different tracks.

What does seem to help is creating space for kids to talk about what they're seeing and how it makes them feel. Not in a way that shames them for caring, but in a way that names the dynamic and makes it less isolating. When a teen can say "I felt bad when I saw that post," and a parent can respond with something like "that makes sense, it's hard to see people having fun without you," the trap loosens a little. The feeling becomes something they can observe rather than something that defines them.

It also helps to model a different relationship with social media yourself. If your child sees you constantly checking your phone, curating your own posts, or commenting on other people's lives, they learn that this is how adults navigate the world. But if they see you putting the phone down, talking about real things, acknowledging your own imperfections, they learn that there's another way to be.

The comparison trap doesn't disappear just because you understand it. But understanding it shifts something. It makes the mechanism visible. And once you can see how it works, you're a little less likely to mistake its whispers for truth.

2.3 Parasocial Relationships

Your daughter talks about a YouTuber the way she talks about her best friend from school. She knows what this creator ate for breakfast, what happened during their recent breakup, how they're feeling about their dog's surgery. She leaves encouraging comments on every video. She worries about them when they go quiet for a few days.

Except they've never met. The creator doesn't know your daughter exists. And yet the relationship feels completely real to her.

This is what psychologists call a parasocial relationship: a one-sided emotional connection where one person invests time, attention, and genuine feeling into someone who can't reciprocate because they don't know the other person is there. The term itself comes from television research in the 1950s, when researchers first noticed that viewers formed attachments to news anchors and sitcom characters. But the internet has transformed what was once a relatively passive experience into something that can feel remarkably intimate and mutual.

Consider what's different now. A child watching a favorite streamer isn't sitting on a couch watching a distant figure on a screen. They're watching someone speak directly to a camera, often in a bedroom that looks like theirs, using the same slang they use, talking about problems they recognize. The streamer says "hey guys" as if greeting friends. They ask questions: "What do you think I should do?" They read comments aloud, sometimes by name. They share behind-the-scenes glimpses, personal struggles, unfiltered moments. The relationship is designed to feel direct and reciprocal, even though it only flows one way.

The production style itself reinforces this feeling. Many successful creators have figured out that authenticity, or at least the appearance of it, builds connection. They film themselves in casual clothes, without scripts, stumbling over words, letting the camera roll during awkward moments. They create ongoing narratives across videos: will they reconcile with their ex, finish renovating their room, get into their dream college? A child tuning in regularly isn't just consuming content. They're following someone's life, rooting for them, feeling invested in outcomes.

And here's what makes this particularly powerful for young people: the relationship can meet real emotional needs. A child who feels misunderstood at home might find a creator who speaks their language. A teen struggling with identity might discover someone who shares their experience and seems confident about it. A kid who feels lonely can visit a creator's channel and find a community of other fans who feel like friends, united by their shared affection for the same person.

The creator, meanwhile, has built a system that encourages exactly this dynamic. They upload on a schedule, training their audience to return. They respond to some comments, making everyone feel seen. They create members-only content, special access, inside jokes that reward loyalty. They share just enough vulnerability to feel human, but not so much that they lose their aspirational quality. They're relatable and extraordinary at once.

What children often don't see is the structure underneath. This isn't a friendship that developed organically between two people. It's a relationship engineered, sometimes quite deliberately, to generate attention and revenue. The creator benefits from every view, every comment, every subscriber. The warmer the parasocial bond, the more reliable the audience. A child who feels personally connected will keep coming back, will defend the creator against criticism, will maybe even spend money on merchandise or memberships.

None of this means the creator is manipulative or malicious. Many are genuinely kind people who care about their audience in a general sense. But the asymmetry remains. The creator wakes up to thousands of messages and can't possibly respond to all of them. They make decisions based on their career, their mental health, their financial needs, not based on any individual viewer's feelings. They might disappear for months, rebrand completely, or say something hurtful without ever knowing a particular child is watching.

For the child on the other side, though, the feelings don't care about asymmetry. If you've spent hours watching someone, if you've laughed with them and worried about them and thought about them when you're doing other things, your brain processes that as a relationship. The emotional experience is real, even if the connection isn't mutual.

This becomes especially tender during adolescence, when young people are naturally pulling away from parents and looking for new figures to admire and learn from. A parasocial relationship can feel safer than a real friendship. There's no risk of rejection, no vulnerability required, no chance the other person will let you down in person. You can feel close to someone without the complicated work of actual intimacy.

Until the illusion cracks. Maybe the creator gets into a scandal. Maybe they retire suddenly, or change in a way that feels like betrayal. Maybe the child tries to make contact and gets nothing back, or worse, a form letter. The grief that follows can be real and surprisingly deep, because the relationship felt real. But there's no closure, no conversation, no mutual acknowledgment of what's ending.

Parents watching this dynamic unfold sometimes feel confused or even a little hurt. Why does my child care so much about this stranger? Why do they get so defensive when I suggest taking a break from these videos? It can help to remember that what looks irrational from the outside reflects something very human: the desire to connect, to belong, to find people who understand you. The medium is new, but the need underneath is ancient.

The question isn't whether these relationships are inherently harmful. For many young people, they're mostly harmless, even occasionally enriching. A kid might discover new interests, find representation they don't see elsewhere, or learn something genuinely useful. The question is whether a child understands what they're experiencing, whether they have other relationships that offer true reciprocity, and whether they're building skills for connection that will serve them when the screen goes dark.

Because the screen always does go dark eventually. Creators move on. Interests shift. What remains is whether a young person has learned to invest in relationships where both people show up, where both people matter, where being known is possible because someone is actually paying attention.

2.4 Escapism as Emotional Regulation

When a teenager comes home from school and immediately disappears into their bedroom with their phone, parents often interpret this as avoidance or antisocial behavior. But something more complex is usually happening. That phone isn't just a distraction. It's a tool for managing an emotional state that feels overwhelming.

Escapism gets treated like a character flaw, but it's actually a form of emotional regulation. We all do it. Adults scroll through news feeds after a frustrating day at work, lose themselves in a novel when anxiety creeps in, or binge a series to avoid thinking about an unresolved conflict. Digital spaces simply make escapism more accessible, more immediate, and more finely tuned to our psychological needs.

The difference for young people is that their capacity to regulate emotions is still developing. The prefrontal cortex, which helps us pause and choose how to respond to stress, won't fully mature until the mid-twenties. In the meantime, kids are navigating academic pressure, social dynamics, identity questions, and a level of uncertainty that can feel relentless. When they turn to screens, they're often looking for the same thing adults seek: relief.

Consider what happens in the moment a child reaches for a device after a hard day. Maybe they had a humiliating experience in gym class, or a friendship is fraying, or they're dreading an upcoming test. The world feels too loud. Their thoughts feel too fast. A game, a video, a chat thread offers something predictable and controllable. The emotional volume gets turned down. The nervous system gets a break.

This isn't inherently harmful. Temporary escape can be restorative. The problem emerges when digital escapism becomes the only strategy a young person knows how to use. If every uncomfortable feeling triggers an immediate reach for the phone, they miss opportunities to build other regulatory skills: sitting with discomfort, talking through a problem, moving their body, or simply letting time pass.

Digital platforms understand this dynamic better than most parents do. They're designed to meet emotional needs with remarkable precision. Feeling lonely? Here's a community that shares your niche interest. Feeling inadequate? Here's content that makes you laugh or feel seen. Feeling bored or restless? Here's an endless stream of novelty. The apps don't just capture attention. They soothe, affirm, and distract in ways that feel almost therapeutic.

But this kind of regulation comes with a cost. Real emotional resilience develops through repetition and variety. A child needs to learn that they can tolerate boredom, sit through sadness, and survive social discomfort. They need to discover that talking to a friend, going for a walk, or even doing nothing for a few minutes can also shift their emotional state. When screens short-circuit that process, those other skills don't get practiced.

It's worth distinguishing between healthy breaks and compulsive avoidance. A kid who plays a game for twenty minutes to decompress before starting homework is using escapism adaptively. A kid who games for three hours every night because it's the only way to stop feeling anxious is relying on a single, increasingly fragile coping mechanism. The second pattern doesn't mean the child is weak or unmotivated. It means they haven't yet built a varied emotional toolkit.

Parents sometimes assume the solution is to remove the screens entirely, but that approach misses the point. If you take away the escape route without addressing what the child is escaping from, you're likely to see the stress show up in other ways: irritability, withdrawal, physical complaints, or a different form of avoidance. The goal isn't to eliminate escapism. It's to help young people develop a range of ways to manage their inner lives.

This requires a shift in how we talk about screen time. Instead of asking "How much is too much?" it can be more useful to ask "What is this doing for my child right now?" If the answer is "helping them recover from a genuinely hard day," that's different from "helping

them avoid something they need to face." One is restoration. The other is a pattern that may need gentle intervention.
Imagine a parent noticing that their child reaches for a device every time they seem upset. Instead of reacting with frustration, they might get curious. What feeling is being managed here? Is it loneliness, overwhelm, anger, boredom, or something else? And are there other ways to address that feeling that might work just as well or better?
Sometimes the answer is surprising. A child who seems glued to social media might actually be using it to feel less alone when they're struggling to connect with peers in person. A child who watches endless YouTube videos might be using them to quiet a mind that races with worry. Understanding the emotional function doesn't mean endorsing unlimited use, but it does change the conversation from restriction to skill-building.
The most effective approach involves creating space for other forms of regulation without shaming the ones that already exist. This might mean helping a child notice their own patterns: "I've noticed you usually go online right after school. Does that help you unwind, or does it sometimes make things harder?" It might mean offering alternatives in a low-pressure way: "Want to shoot hoops for a few minutes before you start your homework?" It might mean addressing the underlying stressor: "You've seemed really stressed about that group project. Want to talk about it?"
What matters most is that young people begin to see emotional regulation as something they have agency over, not something that just happens to them. Screens can be part of that process, but they work best when they're one option among many, not the default response to every uncomfortable feeling.
Over time, the goal is for a child to develop what researchers call emotional flexibility: the ability to match a coping strategy to the situation at hand. Sometimes escape is exactly what's needed. Sometimes what's needed is connection, movement, creativity, or simply the passage of time. Building that flexibility takes years, and it requires patience from the adults who are guiding the process.
The pull toward digital escapism isn't a sign that something is wrong with this generation. It's a sign that they're human, navigating a world that offers an unprecedented number of ways to avoid discomfort. Our job isn't to judge that impulse but to help them expand their

repertoire, so that when life gets hard, they have more than one way to find their way through.

2.5 Early Mental Health Signals

When a child starts spending more time on screens, the changes don't usually announce themselves. There's rarely a moment when a parent thinks, "This is it—this is the problem." Instead, what tends to happen is subtler. A kid who used to chat at dinner gets quiet. Bedtime, which used to take twenty minutes, now stretches to an hour of negotiation. A teenager who loved soccer suddenly finds reasons not to go to practice.

These shifts can feel like normal growing pains, and sometimes they are. But they can also be early signals that something about screen use is affecting a young person's mental health. The difficulty is that these signals often look like other things—stress from school, friendship drama, developmental changes. It takes attention and pattern recognition to notice when screens might be playing a role.

Consider sleep first, because it's one of the most immediate indicators. A child who used to fall asleep easily now lies awake. They're tired during the day but wired at night. When a parent checks in, they might discover the child has been scrolling through videos or chatting with friends well past the time they said goodnight. The blue light from screens suppresses melatonin, which makes falling asleep harder, but the content itself can be just as disruptive. A funny video leads to another, then another. A text conversation turns into a group chat spiral. What looks like a sleep problem might actually be a self-regulation problem, and screens make self-regulation much harder because they're designed to keep attention locked in.

Sleep loss compounds quickly. A kid who's short on sleep becomes irritable, has trouble concentrating, and feels more anxious. Parents sometimes interpret these symptoms as behavioral issues or emotional struggles without recognizing that the root cause might be something as straightforward as staying up too late on a device.

Mood changes are another signal, though they're harder to read. Imagine a child who seems fine most of the time but becomes unexpectedly emotional when asked to put their phone down. Or a teenager who seems anxious or low after spending time online, even though nothing specific seems to have happened. Social media, in particular, has a way of destabilizing mood. A kid scrolls through posts and sees friends at a party they weren't invited to. They post something and check compulsively to see how many likes it gets. They read comments and take them to heart, whether they're kind or cruel. This kind of emotional volatility doesn't always look like sadness or anger. Sometimes it looks like withdrawal. A child stops wanting to talk about their day. They seem distant or preoccupied. When parents ask what's wrong, they say "nothing" and go back to their phone. The phone becomes both the source of distress and the tool for avoiding it, which makes the pattern hard to interrupt.

Then there are the relational signals. A young person who used to enjoy family time now seems annoyed by it. Conversations feel effortful. They're physically present but mentally elsewhere, checking notifications or thinking about what's happening online. Friendships can shift too. A child who had close friends in the neighborhood starts spending less time with them in person and more time talking to people they've never met. Or they have a falling out with a friend that seems to have started or escalated through texts and group chats, where tone gets misread and conflicts spiral quickly.

None of these signals, on their own, necessarily means there's a problem. Kids go through phases. They need privacy and autonomy as they grow. But when several of these patterns appear together—disrupted sleep, mood swings tied to screen use, withdrawal from family or in-person friendships—it's worth considering whether screens are playing a larger role than they should.

The tricky part is that these early signals often don't feel urgent. They feel like small inconveniences or normal adolescent moodiness. Parents adjust. They accommodate. They tell themselves it's just a phase. And sometimes it is. But sometimes these early signals are the leading edge of something more serious, like anxiety, depression, or social isolation, and by the time the problem becomes undeniable, the patterns are much harder to change.

What makes this particularly difficult is that the child often doesn't recognize the connection either. They don't think, "I'm feeling anxious because I spent three hours comparing myself to people on social media." They just feel anxious. They don't realize they're tired because they stayed up scrolling. They just feel tired. The cause and effect are obscured, which means both parents and kids can miss what's actually happening.

The mental health signals worth watching for aren't dramatic. They're ordinary moments that feel slightly off. A kid who used to bounce out of bed now drags themselves up. A teenager who loved drawing hasn't touched their sketchbook in weeks. A child who was excited about a school trip now seems indifferent. These moments are easy to overlook, especially when life is busy and there are a dozen other things demanding attention.

But noticing them matters, because early signals are when change is still relatively easy. A conversation about screen time before bed is simpler than treating chronic insomnia. Helping a child reconnect with an old hobby is easier than addressing entrenched isolation. Recognizing that a mood pattern is linked to social media use is more manageable than responding to a full-blown depressive episode.

The goal isn't to become hypervigilant or to blame every difficult moment on screens. It's to stay observant and curious. When something feels off, it's worth asking whether screens might be part of the picture. Not as the villain in the story, but as one factor among many that shapes how a young person feels and functions. That kind of noticing doesn't require expertise. It just requires paying attention to the child in front of you and trusting that small changes in behavior often mean something.

2.6 Spotting an Emotional Hijack (Parent Guide)

There's a moment many parents recognize but struggle to name. Your daughter closes her laptop after an hour on a social app, and something has shifted. She's restless, maybe irritable. She reaches for her phone within minutes, scrolling with a kind of jittery focus that doesn't look like enjoyment. Or your son finishes a gaming session and seems wired, unable to transition to dinner or homework without friction. He's not being defiant exactly. He just can't seem to settle.

What you're witnessing isn't always a character issue or a discipline problem. Sometimes it's an emotional hijack, a state where the app's design has temporarily overridden your child's ability to self-regulate. The child isn't choosing to be difficult. Their nervous system is responding to stimuli engineered to keep them engaged past the point of comfort.

Understanding what an emotional hijack looks like matters because it changes how you respond. Instead of assuming your child is being willful or screen-addicted, you can recognize when their behavior is being actively shaped by design choices they can't see and didn't consent to.

Start with the transition. How does your child move away from the screen? A healthy ending might include mild reluctance but basic cooperation. An emotional hijack often shows up as disproportionate resistance. Imagine a parent asking their teen to pause a video app for dinner. The teen snaps back with sudden anger, or they agree but keep scrolling, unable to actually stop. This isn't garden-variety teenage attitude. It's a nervous system that's been overstimulated and is now in a low-grade fight-or-flight response.

Watch for the inability to disengage even when the child wants to. Consider a situation where your middle schooler says they're tired and want to stop playing a game, but they keep starting one more round. They might express frustration with themselves, saying they hate the game or that it's boring, yet they can't walk away. This paradox is a hallmark of hijacked engagement. The design is exploiting the gap between what the child wants consciously and what their brain is being prompted to do.

Physical signs offer another window. After heavy app use, does your child seem dysregulated in their body? You might notice shallow breathing, tension in their shoulders, or a flushed face. They might be thirsty or hungry but unable to register those needs. Some kids get a glazed, distant look. Others become hyperactive, physically unable to sit still. These aren't just signs of too much screen time. They're signs that the child's stress response has been activated and hasn't yet calmed.

Mood crashes are telling too. Picture a child who seems fine, even energized, while using an app, but becomes tearful, irritable, or despondent shortly after stopping. The shift can feel abrupt and confusing. What happened? Often, the app delivered a steady stream of small dopamine hits through likes, rewards, notifications, or unpredictable content. When that stream stops, the child experiences a kind of emotional withdrawal. They're not sad about something that happened. They're coming down from a chemically driven high.

Another pattern involves reassurance-seeking or social anxiety that spikes around app use. Imagine a teen who posts a photo and then checks their phone obsessively for the next hour, visibly anxious about responses. Or a child who becomes preoccupied with metrics like follower counts or game rankings, tying their self-worth to numbers that fluctuate by design. The app has inserted itself into their emotional regulation system, and they've lost the ability to feel okay without external validation from the platform.

Time distortion is common too. Ask your child how long they've been on a particular app, and they may genuinely underestimate by half or more. This isn't dishonesty. Apps use autoplay, infinite scroll, and other mechanisms that obscure the passage of time. If your child is frequently surprised by how much time has passed, or if they struggle to remember what they even looked at, that's a sign the app is controlling their attention more than they're controlling the app.

Then there's the content itself. Does your child encounter material that leaves them feeling worse? Maybe they're watching prank videos that normalize cruelty, or they're scrolling through aspirational content that makes their own life feel inadequate. Perhaps they're being fed increasingly extreme versions of something they showed mild interest in, because the algorithm rewards intensity over balance. A hijack doesn't always look like overstimulation. Sometimes it's a gradual shaping of mood and worldview, where the child starts adopting the emotional tone of the content without realizing it.

The compulsion to document or perform for the app is worth noticing. Picture a family outing where your child is more focused on capturing the right photo than experiencing the moment. Or a situation where they want to try something primarily because it will make good content. The app has trained them to see their life as material, and their sense of presence has been outsourced to an audience they may not even know.

Sleep disruption often reveals a hijack that's been going on quietly. If your child has trouble falling asleep, or if they wake up and immediately check their phone, their nervous system may still be responding to stimuli from hours earlier. The app's design doesn't end when the screen goes dark. It lingers in the body's state of arousal.

What makes these patterns hard to spot is that they rarely announce themselves. Your child probably won't say, "I feel emotionally hijacked." They might not have the vocabulary or self-awareness to name what's happening. They may feel vaguely bad or out of control but blame themselves rather than the design. Your role isn't to diagnose or fix them. It's to notice patterns they can't see yet and to gently name what might be happening.

This isn't about demonizing apps or assuming every moment of screen use is harmful. Plenty of app interactions are benign or even positive. But when you see a cluster of these signs, especially if they're new or intensifying, it's worth considering whether the app is doing more than entertaining. It may be actively undermining your child's ability to feel calm, present, and in control of their own attention.

The tricky part is that hijacked behavior can look like other things: fatigue, hunger, developmental moodiness, underlying anxiety. You're not looking for a single smoking gun. You're looking for a pattern that consistently follows app use and that doesn't show up in other contexts. If your child can spend an hour reading or building something or talking with a friend and transition smoothly, but struggles after certain apps, the app is probably the variable that matters.

Spotting an emotional hijack doesn't require technical expertise. It requires paying attention to the texture of your child's experience and trusting what you observe. When the app seems to be steering behavior rather than responding to it, when your child looks less like themselves during and after use, you're seeing something real. And once you see it, you can start having different conversations about what's actually going on.

Chapter 3

The Modern Threat Landscape

"Most digital harm doesn't arrive loudly. It arrives incrementally."

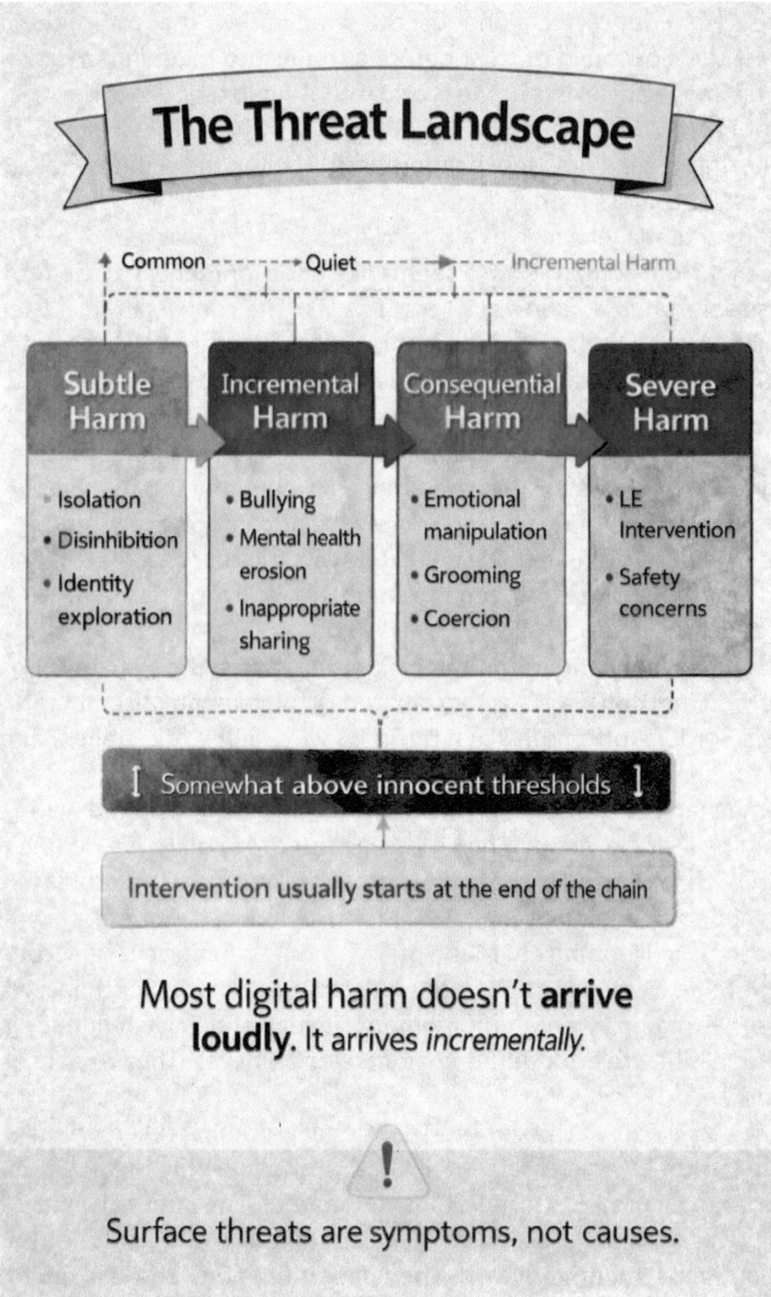
The Threat Landscape
Common
Quiet
Incremental Harm
Subtle Harm
Incremental Harm
Consequential Harm
Severe Harm
• Isolation
• Disinhibition
• Identity exploration
• Bullying
• Mental health erosion
• Inappropriate sharing
• Emotional manipulation
• Grooming
• Coercion
• LE Intervention
• Safety concerns
Somewhat above innocent thresholds
Intervention usually starts at the end of the chain
Most digital harm doesn't **arrive loudly**. It arrives *incrementally.*
Surface threats are symptoms, not causes.

3.1 The Predator Economy

When we think about online predators, most of us picture the warning posters from our own childhoods: the stranger in a chat room, the fake profile, the slow build of trust before a request to meet in person. That mental image still exists in the world, but it no longer describes the threat most kids actually face.

What's happening now looks different. It's faster, more industrial, and built on a business model.

Imagine a teenager receives a direct message on Instagram or Snapchat from someone who seems like another teenager. The profile looks real—photos, followers, recent posts. The conversation starts casually, maybe with a compliment or a shared interest. Within minutes, it turns flirtatious. The other person suggests moving to a more private app, one where messages disappear. They ask for a photo. The teen sends one. Then the mask drops. The person on the other end threatens to send that image to everyone the teen knows unless they pay money or send more explicit content. Sometimes they've already compiled a list of the teen's friends, classmates, and family members. The demand is immediate. The teen has minutes to decide.

This is sextortion, and it's not an isolated tactic used by a handful of disturbed individuals. It's a coordinated criminal enterprise operating across borders, often run like a business with shift work, scripts, and performance targets.

The mechanics are ruthlessly efficient. Criminals don't spend weeks grooming a single victim. They message hundreds of young people at once using fake profiles that cycle through platforms. The profiles are designed to appeal to teenagers: attractive, age-appropriate, algorithmically optimized. Many of the people running these accounts are working from organized crime hubs in West Africa, Southeast Asia, and other regions where enforcement is weak and English fluency is common. They're not operating alone in basements. They're part of teams.

The speed is part of the design. Traditional grooming relied on building trust over time, which meant each predator could only target a limited number of victims. Sextortion flips that model. The criminal's goal is not a relationship. It's a transaction. Get the image, make the threat, collect the payment, move on. The faster it happens, the less time the young person has to think clearly, talk to someone, or recognize the manipulation for what it is.

This industrial approach also explains why sextortion has exploded in the past few years. It scales. A single operation can target thousands of kids simultaneously across multiple platforms. If even a small percentage respond, the payoff justifies the effort. The crime has become so common that some young people now receive these approaches multiple times, from different accounts, using nearly identical scripts.

The financial model is straightforward. Criminals demand payment through apps that are hard to trace—cryptocurrency, gift cards, peer-to-peer payment platforms. Amounts vary, but they're often calculated to be within reach of a panicked teenager: a few hundred dollars, maybe a thousand. Sometimes the demand escalates. Sometimes it stops after one payment. Often, even when the young person complies, the image gets distributed anyway.

But money isn't always the goal. Some sextortion operations are designed to generate more content. The criminal uses the first image to demand additional, more explicit material, which can then be sold, traded in online networks, or used to extort someone else. In this version of the crime, the teenager becomes both victim and product.

It's worth pausing to recognize how different this is from the dangers parents were warned about a generation ago. The threat then was secretive and patient. The threat now is aggressive and automated. It doesn't rely on a child being naive or vulnerable in the traditional sense. It relies on the child being a teenager—impulsive, curious, responsive to attention, and still learning how to assess risk in real time.

Most kids who get caught in sextortion aren't doing anything particularly reckless. They're doing what teenagers have always done: trying to figure out relationships, seeking validation, experimenting with identity. The difference is that those normal developmental behaviors now happen on platforms designed to facilitate instant connection with strangers, and those strangers include people who have systematized the process of exploiting adolescent psychology.

One of the cruelest aspects of sextortion is how it traps young people in silence. The shame is immediate and overwhelming. Many kids believe they've done something irreversibly stupid, that they'll be blamed, or that telling an adult will only make things worse. The criminal counts on this. The threat includes a warning: tell anyone, and we'll release everything. So teenagers drain their savings, borrow money from friends, or try to handle it alone. By the time a parent finds out, the child has often been suffering in isolation for days or weeks.

The global nature of these operations makes them difficult to stop. A criminal in one country targets a child in another, uses servers in a third, and collects payment through a platform in a fourth. Law enforcement agencies are working to coordinate across borders, but the speed and volume of these crimes outpace the resources available to investigate them. Platforms are removing accounts and trying to detect patterns, but new profiles appear as quickly as old ones get shut down.

This isn't a problem that can be solved by teaching kids to "be careful online." The criminals are professionals. They're better at this than most kids are at recognizing it. And they're operating in an environment where the barrier to contact is nearly zero, where profiles can be faked in minutes, and where the design of many platforms actively encourages rapid, unfiltered communication.

What parents are up against, then, is not a scattered collection of threats but a functioning economy. It has supply chains, revenue models, and a workforce that treats this as a job. Understanding that doesn't make the problem easier to solve, but it does clarify what we're dealing with. This isn't about a few bad actors. It's about a system that has learned how to monetize adolescent vulnerability at scale.

3.2 Grooming Pathways and Contact Risk

Most parents think of online predators the way we think of strangers in parking lots: sudden, obvious, and avoidable if you just stay alert. But that's not how it works. The danger isn't a stranger jumping out from behind a screen. It's someone who starts as a friendly teammate, a helpful guild member, or just another kid who likes the same game. The pathway is predictable, and it begins in public.

Imagine your daughter playing an online game where players communicate through team chat. Someone notices she's good at the game. They compliment her strategy. They make a joke that lands. Over several sessions, this person becomes familiar. They're helpful.

They remember details about her gameplay. They're never weird or pushy. They just seem like a friend.

Then comes the shift: "Hey, the voice chat in this game is terrible. Want to move to Discord so we can actually talk during raids?" Or, "I made a private server for people who actually know how to play. Want the link?"

This moment, this invitation to move off the platform, is where grooming accelerates. And most kids don't see it as a warning sign because by then, it doesn't feel like talking to a stranger anymore.

The structure of these pathways is consistent because they exploit something fundamental about how relationships develop. Trust is built slowly, through repeated positive interactions. Predators understand this better than most parents do. They know that after a dozen gaming sessions together, a kid will see them as a known entity. The move to private messaging doesn't feel dangerous. It feels like a natural next step among friends.

Once communication shifts to a private channel, three things change at once. Adult oversight disappears, since most parents don't monitor Discord servers or encrypted messaging apps the way they might glance at an open game chat. The predator gains control over the interaction, without other players present to notice odd behavior or interrupt. And perhaps most importantly, the conversation can gradually shift away from the game itself.

This is where grooming becomes active. The person starts asking questions that feel caring: How's school? What's going on at home? Do your parents understand you? They listen in ways that feel validating, especially to a kid who's going through a rough patch or feels unseen by adults in their life. They remember details. They check in. They become someone the child trusts, sometimes more than they trust the adults around them.

The requests start small. "Send me a photo so I know what you look like." Then, "You should see what I look like too. Here." The boundary between online interaction and real identity begins to blur. By the time overtly sexual content enters the conversation, the predator has often spent weeks or months building enough trust that the child doesn't immediately shut down the interaction. Some kids comply out of confusion. Others do it because they've been convinced this person cares about them, and rejecting the request would mean losing the relationship.

Not every friendly interaction online is dangerous, of course. Kids form real friendships through gaming, and those relationships can be meaningful. The challenge for parents is that the early stages of grooming look identical to normal social connection. There's no moment where a siren goes off. The red flags only become obvious in retrospect.

What makes this particularly difficult is that platforms design their social features to encourage exactly this kind of relationship-building. Games want players to form teams, join guilds, and communicate frequently because it keeps people playing. Moving from in-game chat to Discord or another platform is so common that it's practically expected in many gaming communities. The infrastructure that supports genuine friendship also creates the conditions for exploitation.

The question isn't whether your child should ever talk to people online. That's not realistic, and it's not necessarily desirable. The question is whether they understand that the familiar feeling of friendship doesn't mean someone is safe, and whether they know that certain invitations, no matter how normal they seem, require a pause and a conversation with you.

Some parents hear about grooming and imagine they'll be able to spot a predator by monitoring their child's activity. But you're not looking for someone who seems dangerous. You're looking for someone who seems safe, helpful, and kind until they're not. That's a much harder thing to detect, which is why the most important defense isn't surveillance. It's making sure your child knows the pathway exists, understands why someone might guide them down it, and feels able to come to you the moment something shifts from comfortable to confusing.

3.3 Synthetic Media Threats

A teenager receives a message from someone claiming to have explicit photos of them. The images look real. The background is their bedroom. The face is unmistakably theirs. But the teenager never took those photos. They never sent them to anyone. The images were created using artificial intelligence, and now someone is demanding money to keep them private.

This is not a distant or theoretical problem. The tools that generate fake images, videos, and audio have become accessible to anyone with

a smartphone and a few minutes of time. What used to require technical expertise now happens through apps that cost nothing and require no special knowledge. A person can take a photo from social media, upload it to one of these tools, and produce a synthetic image that looks disturbingly authentic. The same technology exists for voices. A short audio clip, maybe from a video posted online or a voicemail, can be fed into software that replicates someone's speech patterns well enough to fool a listener who isn't paying close attention.
The term "deepfake" originally referred to a specific kind of video manipulation, but it has expanded to cover anything artificially generated that mimics a real person. Photos, videos, voice recordings. The word itself can feel technical or distant, but the impact is immediate and personal. When a young person sees their own face attached to content they never created, the emotional response is visceral. Shame, confusion, anger, fear. It doesn't matter that the content is fabricated. The harm is real.

The scenarios that unfold from here follow predictable patterns. Someone creates a synthetic image or video, then threatens to share it unless the target complies with a demand. Sometimes the demand is financial. Sometimes it's sexual. Sometimes it's simply the act of control itself, the power of making someone feel exposed and vulnerable. The person being targeted often doesn't know where the images came from or how they were made. They only know that something humiliating exists, that it looks like them, and that someone is holding it over them.

What makes this particularly insidious is the way it hijacks normal adolescent fears. Teenagers are already navigating questions about their bodies, their reputations, and how others perceive them. The idea that compromising images might exist, even false ones, taps directly into that anxiety. The threat doesn't have to be carried out to do harm. The existence of the threat is the harm. A young person might comply with demands, withdraw from friends, or spiral into secrecy and shame, all because of something that never actually happened.

Parents sometimes assume that if their child hasn't sent explicit photos, they're safe from this kind of exploitation. That assumption no longer holds. The raw material for synthetic content can come from anywhere: a yearbook photo, a post from a family vacation, a video shared in a group chat. If it's online, it can be used. The boundary between private and public has blurred in ways that make older models of safety and caution insufficient.

There's also a gendered dimension to this. Girls and young women are disproportionately targeted with synthetic sexual imagery. The goal is often humiliation or coercion. Boys are not immune, but the tactics and motivations can differ. A boy might be targeted with fake audio of him saying something offensive, or a manipulated video that makes him look aggressive or out of control. The commonality is the exploitation of reputation and the weaponization of digital identity.

It helps to understand that the people creating and distributing this content are not always strangers. Sometimes they're classmates. Sometimes they're former friends or romantic partners. The tools are so accessible that the barrier between impulse and action has nearly disappeared. A moment of anger or jealousy can lead to the creation of content that follows someone for years. The ease of creation does not correlate with an understanding of consequences. A teenager who generates a fake image might not grasp the legal or emotional weight of what they've done until it's too late.

For parents, this landscape feels overwhelming because it requires vigilance in areas that didn't exist a generation ago. You can't simply tell your child not to send photos and expect that to be sufficient. You have to help them understand that their digital presence, all of it, is potentially material for someone else's manipulation. That conversation is uncomfortable, but it's necessary. It's not about instilling paranoia. It's about building an awareness that the tools of harm have changed, and so must the strategies for protection.

When synthetic content does surface, the response matters enormously. A young person who comes to a parent with this problem needs to be met with calm and belief, not blame or shame. The instinct to ask "What did you do?" is understandable, but it's the wrong question. The question should be "What do you need?" The focus has to be on addressing the immediate threat and mitigating harm, not on assigning fault. If the content is being used for blackmail, that's a crime, and it should be reported. If it's being shared in social circles, that's harassment, and it can be addressed through schools or platforms. The worst outcome is for a young person to carry this burden alone because they're afraid of how adults will react.

There's also the reality that once something is created, it's difficult to fully erase. Even if the original source is removed, copies can circulate.

This doesn't mean the situation is hopeless, but it does mean that expectations need to be realistic. The goal is to limit the spread, hold perpetrators accountable, and support the person who's been targeted. That support might look like therapy, legal action, or simply consistent reassurance that they are not defined by what someone else fabricated.

What's emerging now is a need for a new kind of literacy. Not just about how to use technology, but about how technology can be used against you. Young people need to understand that images can be faked, voices can be cloned, and reputations can be attacked through entirely synthetic means. They also need to know that if they see something shocking or explicit involving someone they know, there's a good chance it's not real. Skepticism is a protective skill. So is empathy. The person in the image or video, real or fake, is still a person, and how we respond to what we see online says something about who we are.

The companies building these tools are not unaware of the harm they enable. Some have introduced safeguards. Some have not. The technology itself is neutral, but the systems that distribute it and the ease with which it's accessed create environments where harm can flourish. As parents, we can't control what tools are available, but we can shape how our children think about them and what to do when they're targeted.

This is not a problem that gets solved with a single conversation or a new app setting. It's part of the ongoing work of raising children in a world where the digital and the physical are increasingly indistinguishable. The threat of synthetic media is real, but so is the possibility of resilience, informed response, and the refusal to let fabricated content define a person's worth or future.

3.4 Peer-Based Harassment and Group Chat Dynamics

The word "bullying" conjures a particular image for many adults: confrontations in hallways, notes passed in class, exclusion at the lunch table. Those things still happen. But much of the cruelty that shapes young people's lives now unfolds in spaces parents rarely see—group chats, shared photo threads, private servers where dozens of kids can participate in something harmful without a single adult knowing it's happening.

The shift matters because the dynamics are different. When harassment moves into private digital channels, it doesn't just become invisible to parents and teachers. It becomes faster, more coordinated, and harder to escape. A comment made in a group chat at 9 p.m. can be screenshotted, shared to three other groups, and have fifty people weighing in before the child who was targeted even wakes up the next morning.

Imagine a middle schooler who makes an awkward comment in a class group chat. Within minutes, someone screenshots it and posts it to a separate chat created specifically to mock them. Twenty kids are added. The original comment gets dissected, exaggerated, remixed into a meme. By the time the child realizes what's happening, the narrative has already solidified. They weren't just awkward—they're now "the weird kid," and everyone has proof.

This is not an accident of technology. It's how group chats function when there's no accountability and when cruelty becomes a form of social currency. The structure itself amplifies harm. In a face-to-face interaction, there are natural limits: physical proximity, the discomfort of saying something mean while looking at someone, the finite number of people who can witness it in real time. In a group chat, those limits vanish. A child can say something cruel from their bedroom, surrounded by the comfort of their own space, while an audience of peers immediately validates the behavior with reactions, replies, or silence that feels like agreement.

The speed also changes how kids process what they're doing. In a physical confrontation, there's usually a moment—however brief—to reconsider. In a group chat, the message is sent before the thought is fully formed. The lack of facial expressions or vocal tone makes it easy to misjudge impact. A sarcastic joke that might have landed as playful teasing in person reads as a genuine insult in text. And once it's out there, it's permanent. Even if someone deletes a message, ten people have already seen it, and three have taken screenshots.

Screenshots are their own form of power. A private conversation between two friends can become public evidence in seconds. A vulnerable confession shared in what felt like a safe space can be used as ammunition days or weeks later. Kids learn quickly that anything they say in a digital space can be weaponized, which creates a climate of distrust even in groups that appear friendly on the surface.

Group chats also enable a particular kind of collective behavior that's harder to sustain in person. When one person makes a mean comment, others feel invited to join in. There's a sense of diffused responsibility—if twenty people are making fun of someone, it feels less like you're doing something wrong and more like you're just participating in what's already happening. The child who might never bully someone one-on-one becomes part of a digital pile-on without thinking of it as bullying at all.

Consider a scenario where a high schooler shares a photo in a group chat and someone responds with a mocking comment about their appearance. A few others add laughing emojis. Someone else screenshots the exchange and shares it to another group with a sarcastic caption. The original poster doesn't know any of this is happening. They just know that people have started acting weird around them at school, making inside jokes they don't understand, looking at them and whispering. The harassment is real, but it's also invisible and impossible to confront because they don't even know where it's coming from.

This creates a particular kind of psychological burden. The child can't identify a single antagonist or a specific moment when things went wrong. There's just a vague, pervasive sense that they've become the subject of something they can't control. They might suspect certain people or certain groups, but they have no proof and no clear path to make it stop. The ambiguity itself becomes part of the harm.

Parents often discover these situations late, if at all. A child might seem withdrawn or anxious, but when asked what's wrong, they say nothing. They're not necessarily hiding it out of defiance. They might genuinely not know how to explain what's happening, or they might assume their parents won't understand the social architecture that makes this situation so painful. Telling an adult feels like admitting weakness, especially if the harassment lives in a space where they're supposed to have autonomy.

The platforms themselves are designed for engagement, not safety. Most messaging apps make it easy to create group chats, add or remove members, and share media, but they offer almost no tools for managing harm. There's no way to flag a conversation as potentially harmful before it escalates. No way to require a cooling-off period before a message is sent. No transparency around who's screenshotting what or where shared content ends up. The architecture assumes good intent, and when that assumption fails, there's little recourse.

It's also worth understanding that many kids don't see this behavior as bullying in the traditional sense. They see it as drama, venting, or just how their social group communicates. The word "bullying" carries weight and moral judgment, and most young people don't think of themselves as bullies. They're just reacting, joking around, being honest. This gap between how adults define harmful behavior and how kids experience it makes intervention tricky. A parent or teacher who steps in with language like "this is bullying" might be met with genuine confusion or defensiveness.

But the harm is real, even when the intent is muddled. A child who becomes the target of group chat harassment often experiences it as relentless and inescapable. They can leave the group, but that doesn't stop people from talking about them. They can block individuals, but screenshots still circulate. They can try to defend themselves, but that often just fuels more mockery. The tools available to them are limited, and the social cost of using them can be high.

What makes this particularly difficult for parents is that the solution isn't as simple as taking away devices or banning group chats. For many kids, these spaces are where essential social life happens. Being excluded from them can mean being excluded from birthday party planning, group project coordination, or just the casual banter that builds friendships. The risk of harm is real, but so is the risk of isolation.

The challenge, then, is helping young people develop the judgment to navigate these spaces without either demonizing the technology or pretending the risks don't exist. That means talking about what cruelty looks like in text form, why screenshots change the stakes of any conversation, and how participation in a pile-on is still participation in harm even if you're not the one who started it. It means creating space for kids to talk about situations that feel confusing or uncomfortable without fear of immediate punishment or device confiscation.

It also means recognizing that the power dynamics in group chats are often invisible to kids themselves until something goes wrong. The child who feels in control one day can become the target the next. The friend group that seems solid can fracture in an evening. Understanding that fragility—and helping kids see it—might be more useful than any rule or restriction.

3.5 Financial Scams and Digital Exploitation

The text message looks harmless enough. A teenager gets an offer to make quick money by rating apps, posting reviews, or "testing" a new platform. Someone slides into their DMs offering cash for feet pics, or promises investment returns that sound too good to be true. What used to be the province of late-night infomercials and email spam has migrated into the apps where teens spend most of their waking hours, and the tactics have evolved to match.

Financial scams targeting young people aren't just about stolen credit card numbers anymore. They're about relationship building, flattery, and exploitation that unfolds over days or weeks. The person on the other end might seem charming, patient, even generous at first. They're not rushing. They're investing.

Take the common pattern that starts with a small task. A stranger reaches out on Instagram or Discord and offers payment for something simple: repost this, rate that, sign up here. The first payment actually arrives. Maybe it's ten dollars, maybe twenty. It feels real because it is real. That's the hook. Trust gets built one Venmo transfer at a a time. Then comes the ask that's a little sketchier—using a parent's payment info "just this once," sharing login credentials, or moving money through a teen's account as part of what's framed as a harmless workaround. By the time the situation turns, the teen is already entangled.

Or consider the romance angle, which is older than the internet but has found new efficiency online. Someone attractive and seemingly older expresses interest. They're complimentary, attentive, and after a while, they mention financial trouble. Maybe they need help with rent, or they're stuck somewhere and need travel money, or there's a family emergency. Sometimes the ask is indirect: they send a payment link and say they'd love to meet up if only they could afford the trip. The emotional investment makes it hard to see what's happening. Saying no feels like abandoning someone who cares about you.

Cryptocurrency and investment scams add another layer because they exploit both greed and ignorance. A teen sees someone online flaunting wealth, showing off gains from trading or NFTs or some new coin. They make it look easy. The pitch is always the same: get in early, don't miss out, this is how people our age are getting ahead. There's a signup link, maybe a referral bonus. Sometimes there's even a real interface that shows fake gains ticking upward. The money looks like it's growing until the moment the teen tries to withdraw it and discovers the whole thing was a facade. By then, any real money they put in is gone.

What makes these scams particularly insidious is how they mimic legitimate opportunities. Teens do make money online. They sell art, edit videos, run social media accounts, trade collectibles. The boundary between real entrepreneurship and exploitation isn't always obvious, especially when the person on the other end is skilled at blurring it. They use the language of empowerment and independence. They position themselves as mentors or allies. They make the teen feel seen and capable, right up until the moment they don't.

There's also the less obvious exploitation that doesn't feel like fraud at first. A stranger offers to pay for photos—nothing explicit, just pictures of hands, or feet, or the teen wearing certain clothes. It starts innocuous and escalates gradually. Or a teen is recruited into a "business opportunity" that turns out to be a pyramid scheme, draining money from friends and family under the guise of hustle culture. These situations don't always trigger alarm bells because they don't start with an obvious villain. They start with someone who seems helpful, even generous.

Parents often don't find out until money is missing, accounts are compromised, or a teen admits they're in trouble. And even when warning signs are there—secretive phone use, sudden interest in money transfers, mentions of online "friends" offering opportunities—it can be hard to intervene without feeling like you're stifling independence or invading privacy. The tension is real. Teens are supposed to be learning how to navigate the world, including its financial dimensions. But the world has come into their bedroom, and it's wearing a very convincing mask.

The challenge isn't teaching teens that all strangers are dangerous or that all online opportunities are fake. It's helping them develop the instinct to pause when something feels off, even if they can't articulate why. It's making sure they know that real opportunities don't require upfront payment, that real relationships don't hinge on financial transactions, and that urgency is almost always a red flag. It's also about creating space for them to come to you when they're unsure, without fear of judgment or a lecture about what they should have known better than to do.

Exploitation thrives in silence. It counts on shame, on the fear of looking foolish, on the worry that admitting a mistake will mean losing freedom or trust. The antidote isn't tighter control. It's an environment where your teen knows they can surface a weird situation and you'll help them think it through, not as a failure on their part, but as a puzzle you're solving together.

3.6 Law Enforcement Realities

When something troubling happens online involving a child, most parents imagine that reporting it will set a clear process in motion. You contact the police, they investigate, and justice follows. But the reality of how digital cases actually work is more complicated, and understanding that complexity can change how you respond in the critical early hours.

Consider a parent who discovers their teenager has been receiving sexually explicit messages from an adult. The instinct is often to delete everything immediately, to scrub the phone clean and protect the child from having to see those messages again. It's a completely understandable impulse. But that deletion can make it nearly impossible to pursue any legal action, even when a crime has clearly occurred.

Digital evidence is fragile in ways that physical evidence is not. When you delete a message from a phone or an app, it doesn't just disappear from your view. In many cases, it's gone entirely. Some platforms keep temporary server logs, but others don't. Some retain data for law enforcement requests, but only for a limited window. The variation is enormous, and investigators often don't know what they'll be able to retrieve until they try.

Imagine a detective receiving a report three weeks after the fact. The messages are gone. The account might be deactivated. The metadata that would show when and where conversations happened has expired from the platform's servers. Without that digital trail, even a willing prosecutor has very little to work with. The case stalls not because anyone doubts what happened, but because there's no longer proof that it did.

This isn't about blaming parents who delete things. It's about recognizing that our emotional response to harm doesn't always align with the requirements of a legal system that operates on documentation and proof. Law enforcement officers who work these cases will often say the same thing: they wish families knew to preserve first and process later.

Preservation doesn't mean your child has to keep looking at harmful content. It means taking screenshots, saving conversations to a separate device, and documenting what happened before anything gets removed. Think of it like photographing a car accident scene before the vehicles are towed. The evidence doesn't stay in your living room, but you need a record of what it looked like when it was.

The timeline matters too. Many parents don't realize that even when evidence is preserved, digital investigations can take months. Subpoenas have to be written and served. Platforms have legal departments that respond on their own schedules. Data has to be analyzed, often by specialists who are managing dozens of other cases. If the person under investigation lives in another state or another country, jurisdiction questions arise. The process is slow, and that slowness can feel like indifference when you're living through it.

There's also a gap between what feels serious to a family and what fits the criteria for criminal charges. A seventeen-year-old sending explicit photos to a fifteen-year-old might be illegal in some jurisdictions and not others. A grown adult sending manipulative but not explicitly sexual messages might fall into a gray area that's hard to prosecute. An anonymous account making threats could be actionable, or it could be impossible to trace. Law enforcement has to work within the boundaries of existing statutes, and those statutes weren't written with modern apps in mind.

This doesn't mean nothing can be done. It means that understanding the system's limitations helps you move through it more effectively. When you know that evidence matters, you document. When you know the process is slow, you find other forms of support while you wait. When you know that some situations won't result in charges, you can still take steps to protect your child and create consequences within your own family or school community.

The officers and prosecutors who handle these cases are often deeply frustrated by the same limitations. They see the harm, they want to help, and they're working within systems that weren't built for the speed and scale of digital life. That frustration doesn't fix the problem, but it does mean that when you interact with law enforcement, you're usually dealing with people who take the issue seriously, even if the outcomes don't always match the gravity of what happened.

One practical reality that surprises many parents: once you make a report, you often lose control over what happens next. If investigators determine that a crime occurred, they may move forward even if your family wants to drop it. If they determine there's insufficient evidence, they may close the case even if you're certain something happened. The legal system operates on its own logic, and families become participants in a process they don't direct.

That loss of control can feel alienating, especially when you're already dealing with a situation that's made your child feel powerless. Some families find it helpful to think of the legal process as one track running parallel to their own healing and safety work. The two tracks don't always sync up, and that's okay. You can cooperate with an investigation while also moving forward with therapy, school interventions, or changes at home. The legal outcome doesn't have to be the measure of whether you've protected your child.

There's also the question of what happens when the person involved is another minor. The legal system treats peer-to-peer situations very differently than adult-to-child cases, and the consequences can range from juvenile diversion programs to serious charges, depending on the circumstances and the jurisdiction. Parents on both sides of these situations often find themselves in bewildering territory, where the language of crime and punishment doesn't quite fit what's actually a developmental issue tangled up with technology.

Understanding how law enforcement operates in digital cases doesn't mean you have to become an expert in digital forensics or criminal procedure. It means recognizing that the system has real constraints, that your response in the first hours can matter enormously, and that legal justice is only one part of how families recover from harm. When you know what to expect, you can make decisions that leave more options open, even if you're not sure yet what you'll want to do with them.

3.7 The Report–Preserve–Escalate Flow

Most parents hit a moment when something online stops feeling manageable at home. Maybe it's a threatening message your daughter received from someone claiming to know where she goes to school. Maybe it's a doctored photo of your son circulating in a group chat.

Maybe it's something you can't quite categorize but that makes your stomach drop.

The instinct in that moment is often to act fast: delete everything, confront the other kid's parents, or call the school immediately. But rushing without a basic framework can make things harder to resolve later. Evidence disappears. Timelines get murky. What felt urgent in the moment becomes difficult to explain a week later when someone in authority asks what actually happened.

The report–preserve–escalate flow is not a legal procedure or an official protocol. It's just a way to think through your next moves when something crosses a line. Think of it as a mental checklist that helps you stay clear-headed when emotions are running high.

Start with preservation, even if you're not sure yet

When you see something troubling, your first move is almost always to document it. Take a screenshot. If it's a video, screen-record it. If it's a voice message, save the file. Capture the username, the timestamp, and any context around it—like the group name or how your child came across it.

This doesn't mean you're filing a police report. It just means you're creating a record before it vanishes. Most platforms allow users to delete posts, unsend messages, or remove accounts entirely. Even if the person who sent the content doesn't take it down themselves, apps sometimes auto-delete after a set period. Once it's gone, it's gone.

Think of this like taking a photo of a dent in your car before you drive it to the mechanic. You're not assuming you'll sue anyone. You just want a clear record of what you saw, in case it matters later.

If your child is shaken or scared, preservation also buys you time. You don't need to decide in that moment whether this is something the school needs to see or whether it rises to the level of law enforcement. You just need to make sure the evidence still exists tomorrow when you've had a chance to think.

Reporting to the platform

Most social apps and games have a way to report content or accounts that violate their rules. This might feel pointless—like shouting into a void—but it's worth doing anyway, especially if the content involves threats, sexual material, impersonation, or harassment.

Platform reports create a trail. If the same account is reported by multiple people, or if the behavior escalates, that history can matter. In some cases, platforms will remove content or suspend accounts quickly. In others, nothing visible happens. But the act of reporting still registers the incident in a system that might be reviewed later by someone with more authority than a front-line content moderator.

You can usually report without your child's account being identified to the other person. The process varies by app, but it typically involves tapping a few buttons on the post or profile itself. If you're not sure how to do it, searching "how to report on [app name]" will get you step-by-step instructions.

Reporting to a platform doesn't replace other actions. It's just one layer. If the situation is serious, you'll probably need to do more.

When to bring in the school

Schools get involved when the behavior affects your child's ability to learn, feel safe at school, or participate in school-related activities. That's true even if the incident happened entirely online and outside school hours.

Imagine your child is in a group chat with classmates where another student threatens to "get him" tomorrow at lunch. Or a photo taken during gym class gets altered and shared with cruel captions. Or kids are using a shared online game to organize exclusion or pile-ons that carry over into the cafeteria. These situations happen off school property and through private apps, but they have a direct impact on the school day.

When you contact the school, bring your preserved evidence. Forward the screenshots, explain the timeline, and describe how it's affecting your child. Be specific: "He doesn't want to go to school tomorrow because of this message" is more useful than "He's upset."

Schools have limited authority over what students do on their personal devices at home. But they do have a responsibility to maintain a safe environment on campus. Most schools now have policies that address cyberbullying or digital harassment, even when it originates outside school walls. They may be able to intervene by talking with the other student's parents, adjusting schedules to reduce contact, or applying disciplinary measures if the behavior violates the student code of conduct.

Not every troubling online interaction requires school involvement. If kids are working something out on their own, or if the issue is entirely outside the school's orbit, bringing administrators in might overcomplicate things. But if your child is losing sleep, refusing to attend, or facing ongoing harassment from classmates, the school is usually the right next step.

When to escalate to law enforcement

Calling the police feels big, and it is. But some situations genuinely require it.

Consider a scenario where your child receives explicit images of another minor, or someone is attempting to coerce her into sharing images of herself. Consider a situation where an adult is contacting your child under false pretenses and trying to arrange a meeting. Consider threats of imminent physical violence, stalking behavior, or blackmail attempts. These aren't school discipline matters. They're crimes.

If you're on the fence about whether something crosses into criminal territory, you can often call a non-emergency police line and describe the situation without filing a formal report. Many departments now have officers who specialize in cybercrimes or crimes against children. They can help you understand whether what you're seeing meets a legal threshold.

When you do make a report, having preserved evidence makes the process much smoother. Officers will want to see exactly what was said, when it was said, and by whom. If you've already taken screenshots with timestamps and saved relevant messages, you're giving them something concrete to work with.

Parents sometimes worry that involving police will make things worse—that it will humiliate their child or escalate conflict with another family. Those concerns are understandable. But in cases involving real harm or the risk of harm, waiting usually doesn't make things better. It just gives the situation more time to unfold in dangerous ways.

The flow isn't always linear

You might report to a platform and realize a day later that you also need to contact the school. You might loop in the school first, then discover details that make a police report necessary. The point of the framework isn't to lock you into a sequence, but to help you think clearly about who needs to know what, and when.

Some situations resolve at the first level. A platform removes the content, or the behavior stops, and nothing else is needed. Others require moving through all three. The key is knowing that you have

options, and that taking one step doesn't prevent you from taking another if the situation changes.
What matters most is that you don't let uncertainty freeze you. Preserve what you can see. Report where it makes sense. Escalate when the situation demands it. You're not overreacting by taking something seriously. You're just making sure that if this turns out to matter, you'll have what you need to protect your child.

FIELD NOTE: The difference between being heard and being turned away
A lot of parents do the right thing emotionally—act fast—but the wrong thing operationally. They go to a large police department with a gut feeling and a vague accusation: "I think something happened to my child."

The hard truth is that big systems run on triage. The first questions are predictable: What happened? Do you have evidence? When did it occur? Where did it occur? Who is involved? What platform or device? Without those basics, families can leave feeling dismissed—when the reality is the report didn't contain enough actionable detail to open the right door.

That means parents sometimes need a hard conversation first. Not an interrogation. A calm, direct fact-gathering talk: what happened, what was said, where it occurred, and what proof exists (messages, usernames, screenshots, dates).

When you can bring a clear timeline and specific evidence, the outcome changes. You get routed correctly, faster. Fear is normal. But clarity is what makes systems move.

Before you report, write down:

- Child's account username(s) + platform(s)
- Offender username(s) / identifiers (if known)
- What happened (one paragraph)
- Timeline (date/time range)
- Where it occurred (online + physical location if relevant)
- Evidence list (screenshots, URLs, message threads)
- Safety status (ongoing contact? threats? meet-up attempt?)

Chapter 4

Behavioral Engineering:
How Tech Is Designed To Win

"Design isn't neutral. It always nudges someone toward something."

Common Dark Patterns

Pattern	Tactic	Manipulation
	Endless Content	Goal becomes to **slow** consumption, not gain benefit.
	Confirmshaming	Make "no" option feel guilt-ridden or foolish.
	Privacy Zuckering	Settings designed to blur what is visible or shared.
	Forced Continuity	Auto-renewal obscured; cancelation process long.
	Nagging	Prevents disengagement through constant nudges.
	False Urgency	"Only 1 left!" / "Sale Ends Soon!" / FOMO tactics.

Design isn't neutral. It always nudges someone toward something. Limit access to frictionless experiences.
Friction is safety.

4.1 The Attention Business Model

When you open most apps your kids use, something invisible happens before the screen even finishes loading. A calculation begins. Not about what your child wants to see, but about how long the app can keep them there.

This isn't a conspiracy. It's a business model, and it works the same way whether the app sells athletic shoes or hosts dance videos. The longer someone stays, the more ads they see, the more data gets collected, and the more revenue flows in. Time equals money, which means your child's attention has a dollar value attached to it.

Picture a shopping mall designed by someone who gets paid every time you walk past another store. The corridors would curve just so. The lighting would draw your eye. Exit signs would be small and poorly marked. You'd find exactly what you came for, sure, but only after passing dozens of other temptations. That's not an accident, and neither is the design of the apps on your child's phone.

The technical term is "engagement optimization," but the simpler truth is this: platforms need to be sticky. They need people coming back, staying longer, and checking more often. Every feature gets tested against that goal. Does this color scheme keep people scrolling? Does this notification timing bring them back faster? Does this video recommendation lead to three more videos, or just one?

Consider what happens when a platform decides whether to show your child a "you've been on for 30 minutes" reminder. That reminder costs engagement. Some users will close the app. The platform knows this because they've measured it. So the decision isn't really about what helps the user. It's about how much engagement they're willing to sacrifice to avoid criticism.

This creates a kind of architectural tension in every app your child uses. The platform wants to be helpful enough that people return, but not so helpful that they leave satisfied. A teenager who finds exactly what they're looking for in two minutes and then puts their phone down is, from the platform's perspective, a failure. The goal is more like fifteen minutes, then twenty, then an hour.

The mechanics of this show up in small ways that compound. Infinite scroll means there's no natural stopping point, no bottom of the page where you'd normally decide you're done. Autoplay means the next video starts before you've chosen to watch it, using inertia instead of interest. Streaks and badges create artificial reasons to return daily, even when you have nothing you actually want to do there.

None of this requires malice. The people designing these systems aren't sitting around plotting how to harm kids. They're solving the problem they've been given: maximize engagement within whatever constraints exist. If a feature increases time spent by three percent without triggering too many complaints, it gets built. If it decreases time spent, it doesn't. The system optimizes itself.

What makes this particularly complicated for families is that engagement optimization often does create real value. A platform that keeps your child watching educational content for thirty minutes has arguably done something good. The algorithm that recommends the perfect song at the perfect moment feels like magic. The app that reminds your teenager to check in with a friend might strengthen that relationship.

But the business model doesn't distinguish between good engagement and empty engagement. It can't. A minute of your child zoning out watching someone else play video games counts exactly the same as a minute of them learning guitar. The platform has no financial incentive to prefer one over the other, and often the zoning out produces more ad views.

This is where the design choices start to matter. Imagine a platform deciding whether to show a "take a break" prompt after continuous use. They test it. Some users appreciate the nudge and feel better about the app. Other users find it annoying and switch to a competitor. The platform now faces a choice: prioritize user wellbeing or prioritize market share. The answer usually depends on whether they think regulators are watching.

You can see this tension in how platforms talk about these issues. They'll announce new wellbeing features with press releases and blog posts, then bury those features three menus deep where hardly anyone will find them. They'll create time limit tools that are easy to dismiss or extend. They'll default to the most engagement-friendly settings and make the calmer ones opt-in. These aren't oversights. They're the business model asserting itself.

For parents, understanding this helps make sense of things that otherwise seem baffling. Why does the app keep suggesting one more video when your child is clearly tired? Because tired people are still engaged people. Why does turning off notifications require six taps but turning them on takes one? Because more notifications mean more returns, and more returns mean more revenue. Why does the app make it so easy to skip past age verification? Because underage users represent growth.

The important thing to recognize is that your family's goals and the platform's goals aren't aligned. You want your child to use technology in ways that support their development, relationships, and wellbeing. The platform wants your child to use technology as much as possible, in whatever way creates the most engagement. Sometimes those goals overlap. Often they don't.

This doesn't mean these platforms are unusable or that every feature is suspect. But it does mean that waiting for platforms to prioritize your child's wellbeing over their own growth is probably not a workable strategy. The business model is too strong. The incentives point in one direction, and thoughtful parenting often requires pointing in another.

4.2 Dark UX Patterns

You open an app to check one thing. Twenty minutes later, you're still scrolling. You meant to glance at a notification. Now you're three videos deep into content you didn't search for. This isn't an accident, and it's not a failure of willpower. The apps your kids use are designed to create these moments.

Dark UX patterns are interface choices that guide users toward actions that benefit the platform, often at the expense of what the user actually wanted to do. The term "dark" doesn't mean evil in a cartoon villain sense. It means obscured. These patterns make it harder to see what's happening or to act on your own intentions. They work on adults, and they work even better on children, whose impulse control and decision-making systems are still under construction.

Some of these patterns are obvious once you know to look for them. Others are subtle enough that even adults don't consciously register what's happening. Imagine your daughter opens a social app to post a photo. Before she can do that, the app shows her a grid of new posts from people she follows, several of which have comments or reactions she hasn't seen yet. What was a simple task now has four or five decision points, each one pulling her deeper into the app. She might still post her photo. But she's already spent five minutes doing other things first, and the app has successfully extended her session.

The infinite scroll is probably the most recognized dark pattern. There's no natural stopping point, no bottom of the page. The content just keeps coming, often with a small loading animation that creates a tiny moment of anticipation. For a child who's still learning to manage time and recognize their own fatigue, this design makes it genuinely difficult to put the device down. It's not that they're ignoring you when you ask them to stop. It's that the interface has removed the environmental cues that usually signal an endpoint.

Autoplay works in a similar way. One video ends, and the next one starts within seconds. The friction of choosing whether to continue has been eliminated. Instead, stopping requires an active decision. Your son has to interrupt his own experience, override the momentum the app has created. That's a harder cognitive task than it sounds like, especially for a young person whose prefrontal cortex is still developing.

Some patterns operate on social pressure rather than momentum. Snapstreaks, for example, require users to send messages back and forth with friends every single day to maintain a streak counter. Missing a day means losing the streak. What looks like a fun feature is actually a commitment device that turns friendship into a daily obligation. Kids will describe feeling anxious about losing streaks, sometimes prioritizing them over homework or sleep. The design has turned communication into a metric and made that metric feel high-stakes.

Read receipts and typing indicators create a different kind of pressure. When your child can see that a friend has read their message, or is typing a response, the interaction starts to feel synchronous even though it's happening through text. Waiting for a reply becomes more stressful. Not replying quickly feels ruder. These small interface details change the emotional texture of communication, making it harder for kids to set boundaries around when and how they respond.

Then there are the patterns that exploit incomplete understanding. Imagine a game that offers your son a "special limited-time offer" on in-game currency. The interface shows a timer counting down and uses visual language that suggests urgency and exclusivity. What it doesn't show clearly is that similar offers appear regularly, or that the discount isn't actually much of a deal compared to standard pricing. The design is counting on his inexperience as a consumer and his still-developing ability to evaluate value.

Defaults matter too. Many apps are set up so that the most privacy-invasive option is the default, and choosing something more restrictive requires navigating through settings menus. Location tracking might be on unless you turn it off. Contact syncing might happen automatically. These aren't technically forced choices, but they rely on inertia and the fact that most users, especially young ones, never dig into settings at all.

Some patterns are built around variable rewards, the same psychological mechanism that makes slot machines compelling. Your daughter posts something and doesn't know if she'll get five likes or fifty, or when they'll arrive. Each time she checks, there's a small possibility of a positive surprise. That uncertainty is more engaging than a predictable outcome would be. Her brain starts treating each check as a potential reward, and the behavior becomes harder to resist over time.

Notifications are their own category of dark pattern, not because they exist but because of how they're designed. Many apps send far more notifications than necessary, and they're often vague enough to require opening the app to understand what happened. "You have new activity" or "Someone mentioned you" creates curiosity without providing information. The notification isn't really informing your child. It's creating an itch that can only be scratched by opening the app, at which point other design patterns take over.

What makes these patterns particularly concerning with children is that they're encountering them without the context adults have. You've probably developed some immunity to manipulative design through years of experience. You've learned to recognize when you're being nudged. Your child is building that literacy from scratch, in an environment that's far more sophisticated than anything you faced at their age.

Understanding dark patterns doesn't mean you need to eliminate every app that uses them. That's not realistic, and it might not even be desirable. But it does mean you can start conversations with your child that go beyond "you're on that thing too much." You can talk about how these interfaces work, what they're designed to do, and how that design might be affecting the experience of using them. You can help your child start to notice when they're being guided toward something they didn't choose.

It also means you can make more informed decisions about which apps and platforms to allow, and under what conditions. An app that uses infinite scroll might be fine for limited, supervised use. An app that combines infinite scroll with autoplay, social pressure features, and aggressive notifications might be worth avoiding until your child is older. The question isn't whether an app uses any manipulative design, because most of them do. The question is how many of these patterns are layered together and how intense their effects are.

The interface is never neutral. Every choice a designer makes about how an app works is also a choice about how the user will experience time, attention, and autonomy while using it. Recognizing those choices is the first step toward helping your child develop a more intentional relationship with the screens in their life.

4.3 The Loop of Obligation

When a twelve-year-old checks Snapchat before brushing her teeth in the morning, she's not just being careless. She's responding to a small, persistent pressure that doesn't announce itself as pressure at all. The app shows her that she's maintained a 47-day streak with her best friend. That number glows quietly on her screen. It carries weight.

Streaks feel harmless. They're playful, even. But they function as a kind of social contract that kids didn't exactly sign up for. The streak creates a mutual expectation: if you care about me, you'll send something every day. If I care about you, I'll do the same. Missing a day doesn't just erase a number. It sends a message, or at least feels like it does.

The same dynamic shows up in group chats where read receipts are visible. Imagine a teenager sees that six friends have all read his message. Five replied within minutes. One didn't. That silence becomes loud. He starts to wonder if he said something wrong, if he's being left out, if the friendship is shifting. Meanwhile, the friend who didn't reply was simply in class with her phone in her backpack. But the design of the platform turned her ordinary absence into something that feels like a choice.

This is what obligation looks like when it's been gamified. The platform creates a structure that nudges kids toward constant participation, not through explicit rules but through social mechanisms that feel personal. The streak isn't enforced by the app. It's enforced by the relationship itself, or by what the child imagines the relationship requires.

Badges and achievement systems work in a slightly different way. When a learning app awards a badge for logging in seven days in a row, it's borrowing the language of accomplishment. The badge suggests progress, mastery, commitment. But what it actually measures is compliance with a login schedule. Kids begin to associate the badge with success, even when the activity itself hasn't deepened their understanding of anything. The loop tightens: open the app, get the reward, feel the small glow of achievement, repeat tomorrow to avoid losing it.

The design here is intentional, though not necessarily malicious. Platforms want users to return. Consistent engagement improves metrics, builds habits, keeps people inside the ecosystem. For adults, this might mean checking email more often than necessary. For kids, whose sense of social belonging is still forming, the stakes feel higher. A missed streak can seem like a failed friendship. A notification ignored might feel like a social error.

Parents sometimes misread this behavior as addiction, but that's not quite right. A child who checks her phone compulsively at dinner isn't necessarily hooked on the content. She might be responding to a steady hum of social obligation that the platform has made visible. Before smartphones, friendships carried unspoken expectations too. But those expectations had natural limits. You couldn't see whether your friend had read your note. You didn't lose a numbered streak if you didn't talk for a weekend.

Digital platforms remove those buffers. They turn informal social rhythms into quantified, visible systems. And because kids are still learning how relationships work, they often take these systems at face value. If the app says the streak matters, it must matter. If everyone else is responding instantly, maybe that's the standard.

The tricky part for parents is that these loops don't always look problematic from the outside. A child might seem engaged, connected, even responsible. She's keeping up with her friends, maintaining her streaks, responding promptly. But underneath, she may be operating on a treadmill she didn't choose, feeling responsible for maintaining a rhythm the platform designed.

Some families have found that simply naming this dynamic can loosen its grip. When a parent says, "It sounds like that streak is starting to feel like a job," the child might realize she's been treating it that way. The obligation becomes visible as an obligation, not just as what friends do. That shift in perspective doesn't solve everything, but it creates a little breathing room.

Platforms will keep designing these loops because they work. They keep users returning, and returning users make platforms viable. The question isn't whether these systems exist. It's whether kids understand that the urgency they feel is often something the design is creating, not something the friendship actually requires.

Friendships can survive a broken streak. Messages can go unread for a few hours without meaning anything. But kids need help seeing that, especially when the design of the platform keeps suggesting otherwise.

4.4 Healthy vs. Manipulative Design

When your child opens an app, something happens in the first few seconds that you probably don't notice. The screen loads in a particular way. Colors appear. A notification might ping. A character might wave hello. These aren't accidents. Every pixel, every sound, every tiny delay has been chosen by someone who understands how human attention works.

Some of these choices respect your child. Others don't.

The difference isn't always obvious, and that's the problem. An app can look cheerful and innocent while using the same psychological tricks that keep adults scrolling past midnight. Your seven-year-old doesn't know she's being manipulated. She just knows she really, really wants to open that app again.

So how do you tell the difference?

What healthy design actually does

Think about a well-made educational app your child has used, one that actually taught them something. Maybe it was a reading app or a math game. When they finished a level, the app probably celebrated their success, then gently suggested they try the next challenge or take a break. It didn't beg them to stay. It didn't make them feel like they were missing out by leaving.

Healthy design has a goal beyond keeping your child's eyes on the screen. It wants to teach something, facilitate creativity, or help them connect with someone they already know in real life. The app becomes a tool for something your child wants to accomplish, not an end in itself.

When an app is designed well, your child can stop using it without feeling anxious or incomplete. They might be excited to come back later, but they're not compelled. There's a natural rhythm to their use. They do something, they finish, they move on.

The app also makes its rules clear. Your child understands what they're working toward and when they've achieved it. There are no sudden changes to how things work, no surprise costs or locked features that appear out of nowhere. The app keeps its promises.

What manipulation looks like

Now imagine an app that never quite lets your child feel finished. Every time they complete something, a new notification appears. A timer starts counting down. A friend's avatar shows up with a gift they can only claim in the next ten minutes. The app creates a sense of urgency where none needs to exist.

This is manipulation, and it works by making your child feel like they're always on the edge of missing something important. The app might use bright red badges that trigger a low-level anxiety until they're cleared. It might send notifications about events that aren't really events, just ordinary features dressed up as limited-time opportunities.

Some apps go further. They introduce streaks that your child worked hard to maintain, then send push notifications when that streak is in danger. They show exactly how many friends are online right now, making your child feel isolated if they're not there too. They use countdown timers on features that will actually be available tomorrow with a different label.

The goal isn't to help your child do anything. The goal is to make sure your child comes back as often as possible and stays as long as possible, whether or not that serves them.

The emotional tells

One reliable way to spot manipulation is to watch what happens when your child tries to stop using an app. Do they close it easily, or does the app throw up obstacles? Some apps make you tap through multiple "Are you sure?" screens. Others show you everything you'll miss if you go. A few even use sad faces or disappointed characters to make your child feel guilty for leaving.

Pay attention to your child's mood after using an app too. Healthy design tends to leave kids feeling accomplished, calm, or energized in a positive way. Manipulative design often leaves them irritable, restless, or vaguely dissatisfied. They might immediately ask to use the app again, not because they enjoyed it but because it left them wanting something they can't quite name.

That empty feeling is a feature, not a bug. Apps designed to maximize engagement often provide just enough reward to keep your child returning, but never quite enough to feel truly satisfied.

The language of dark patterns

Designers have a term for manipulative techniques: dark patterns. These are interface choices that trick people into doing things they didn't mean to do, or that make it unreasonably hard to do what they actually want.

In kids' apps, dark patterns show up in predictable ways. A button that looks like it will let your child skip a commercial actually opens the app store. The "X" to close an ad is so small that your child keeps missing it and clicking the ad instead. The app asks for permission to send notifications and makes "yes" look like the obvious choice while "no" is small, gray, and apologetic.

Sometimes the manipulation is more subtle. An app might let your child play the first few levels for free, then introduce a feature that requires payment or viewing ads, but frame it as though your child is choosing to unlock a bonus rather than being asked to pay for something that used to be free.

These patterns rely on your child not understanding what's really being asked of them. The app takes advantage of their trust.

“If stopping feels uncomfortable, that’s a design outcome—not a personal failure.”

When "free" gets complicated

This is worth looking at more closely because it confuses adults too. Many apps are free to download but make money by showing ads, selling in-app purchases, or collecting data. None of these business models are inherently wrong, but they create incentives that don't always align with your child's wellbeing.

An app that makes money from ads wants your child to see as many ads as possible, which means keeping them in the app as long as possible. An app that makes money from in-app purchases wants your child to hit a frustration point where paying seems like the only way forward. An app that makes money from data wants to learn as much as possible about your child's behavior, preferences, and patterns. Healthy apps handle these tensions transparently. They might show ads but not interrupt your child in the middle of an activity. They might offer purchases but make sure the free version is genuinely playable. They might collect usage data but explain why and let you opt out. Manipulative apps hide these tensions or exploit them. They make the free version so frustrating that it barely functions. They place ads at moments of high emotion or concentration. They nudge children toward purchases without making it clear that real money is involved.

What you can do with this

You don't need to become a design expert to use this information. You just need to start noticing patterns. Spend a few minutes watching your child use an app without saying anything. See what the app does when your child tries to leave. Notice whether your child seems in control of their experience or whether the app is steering them somewhere.

Ask your child questions about what's happening on screen. "Why did that pop up just now?" or "What does that timer mean?" Sometimes they know they're being pushed but don't have the language to express it. Other times they haven't noticed at all, and your question helps them see it.

When you find an app that treats your child well, that becomes your benchmark. You'll start recognizing the difference more quickly because you'll know what respect looks like in an interface.

The truth is, most apps contain some mix of healthy and manipulative design. Very few are purely one or the other. What matters is whether the app's overall approach respects your child's autonomy and wellbeing, or whether it treats your child's attention as a resource to extract. Once you learn to see that distinction, it's hard to unsee.

4.5 Counter-Conditioning the Brain

Your daughter reaches for her phone the moment she sits down at the breakfast table. Not because anything happened. Not because she's expecting a message. Just because the phone is there, and her hand moves toward it the way your hand moves toward a seatbelt when you get in the car.

This isn't a character flaw. It's conditioning. Her brain has learned, through thousands of repetitions, that unlocking the screen delivers a small burst of possibility. Maybe a notification. Maybe a funny video. Maybe nothing at all. But the maybe is enough to keep the loop running.

The designers who built these apps understand conditioning better than most parents do. They've borrowed principles from behavioral psychology—the same ones used to train animals, treat phobias, and help people quit smoking—and deployed them to make their products feel irresistible. The good news is that conditioning works both ways. If a habit can be built through repetition and reward, it can also be weakened, rerouted, or replaced.

Counter-conditioning doesn't mean forcing your child to go cold turkey or lecturing them about dopamine. It means quietly introducing friction, alternate rewards, and new patterns until the old autopilot response begins to fade. Think of it as retraining the reflex rather than fighting the urge.

Start with the physical environment. If the phone lives on the nightstand, it becomes the first thing your child sees in the morning and the last thing they touch at night. That's two powerful conditioning moments every single day. Moving the phone to a drawer across the room, or better yet, to a shared charging station in a common area, breaks the spatial association. The phone is no longer an extension of the bed. It's a tool that lives somewhere else.

This kind of change sounds trivial until you try it. The first few nights, your child will probably feel the absence. They might say they need the phone for an alarm. Get them an alarm clock. They might say they need it in case of emergency. Remind them that for most of human history, children slept without a phone within arm's reach and emergencies were handled just fine. You're not being unreasonable. You're weakening a conditioned response by removing the cue.

The same principle applies to mealtimes. Phones at the table create a conditioned expectation that eating and scrolling go together. If you want to counter that, the phone has to be absent—not just silent, but out of sight. Some families use a basket by the door. Some use a drawer. The method matters less than the consistency. When the phone isn't there, the brain stops reaching for it during meals. The conditioning weakens.

Another lever is notification settings. Every ping is a little jolt of attention, and over time, those jolts train the brain to stay on alert. Turning off all non-essential notifications—so the phone only interrupts for actual people trying to reach your child, not apps trying to pull them back in—removes hundreds of conditioning events per week. It won't feel like much at first. But after a few weeks, you might notice your child checking their phone less often, simply because the cues have gone quiet.

Grayscale mode is another surprisingly effective intervention. Most phones have a setting that drains all the color from the screen. It doesn't block anything. It doesn't limit access. It just makes everything look dull. Colors are rewarding to the brain. They signal novelty and excitement. A grayscale screen signals the opposite. Videos look less appealing. Photos feel flat. The apps still work, but they've lost some of their pull. This is counter-conditioning at the sensory level.

Some parents worry that these strategies feel manipulative, like they're sneaking around instead of having honest conversations. But you're not hiding anything. You can explain exactly what you're doing and why. You can say, "These apps are designed to grab your attention in ways that make it hard to stop. We're changing a few things to make it easier for your brain to take a break." That's not manipulation. That's teaching your child how design works and helping them reclaim some control.

Counter-conditioning also means creating alternate patterns that compete with the phone. If your child is used to scrolling while waiting for dinner, try handing them a task instead. Set the table. Chop vegetables. Fold napkins. If they're used to reaching for their phone the moment they get in the car, try starting a conversation or playing a song before they have a chance to pull it out. You're not policing their behavior. You're offering their brain a different option before the old habit kicks in.

This works because the brain doesn't care much about what the habit is, only that there's a clear cue, a routine, and a reward. If you can insert a new routine between the cue and the reward—or better yet, remove the cue entirely—the old pattern begins to lose its grip.

Imagine a family that starts a ritual where everyone's phone went into a bowl on the kitchen counter from dinner until bedtime. No one was allowed to check it during that window unless they could name a specific reason. The first week was hard. The kids kept drifting toward the bowl. But by the second week, the pull had softened. By the third week, it felt normal. The conditioning had shifted. The evening became a phone-free zone not because of willpower, but because the environment had changed and the brain had adjusted.

That's what counter-conditioning looks like in practice. Small changes, repeated consistently, until the automatic behavior is no longer automatic.

It's worth saying that none of this will work if the rest of your child's life feels hollow. If school is stressful, friendships are strained, and there's nothing else that feels good, the phone will keep winning. Counter-conditioning creates space, but that space has to be filled with something—connection, rest, boredom, creativity, physical activity, face-to-face time with people they care about. The brain is always learning. The question is what you're teaching it.

4.6 The 48-Hour Notification Detox

Most of us don't realize how much of our attention has been auctioned off until we try to buy it back.

Notifications feel like they're serving us. They promise to keep us connected, informed, on top of things. But in practice, they've trained us to respond to our phones the way Pavlov's dogs responded to a bell. A buzz, a banner, a red dot, and suddenly we're reaching for the device without deciding to. Our kids watch this happen dozens of times a day. They see us mid-sentence, mid-meal, mid-story, and then gone for a moment, our attention yanked somewhere else. And they're learning to live the same way.

This subsection describes a short experiment that can help your family feel what it's like to own your attention again, even briefly. It's not about punishing anyone or proving a point. It's about noticing what changes when the interruptions stop.

The basic idea is simple. For 48 hours, you turn off all non-essential notifications on your phone and on your child's device. Not the phone itself. Not the apps. Just the alerts, the pings, the things that reach out and grab you. You still have access to everything. You just have to choose to look rather than being told to look.

What counts as non-essential depends on your life, but here's a reasonable starting place. Keep notifications for phone calls, text messages from real people, and calendar reminders if those help you get places on time. Turn off everything else. Social media. Games. Email. News. Delivery updates. App suggestions. Marketing messages.

Likes, comments, shares, streaks, all of it. If it's designed to pull your eyes to the screen, it's off for two days.

Before you begin, talk with your child about what you're both trying. Explain it as an experiment, not a punishment. You might say you've noticed that both of you check your phones a lot when the alerts go off, and you want to see what it feels like to only check when you actually want to. Emphasize that nothing is being taken away. Every app still works. You're just changing how they call for attention.

Then help them turn the notifications off. Sit with them, go through the settings together. This matters for two reasons. First, it's surprisingly complicated to silence everything, and your child may not know how. Second, doing it together makes it a shared project rather than something being done to them.

For the next two days, pay attention to what happens. You'll probably reach for your phone out of habit and find nothing waiting for you. That moment of silence can feel strange at first, almost like something's wrong. You might even feel a small spike of anxiety, a worry that you're missing something urgent. That feeling is worth noticing. It tells you how much your nervous system has been shaped by constant alerts.

Your child will likely have a stronger reaction. They've grown up in this environment. The absence of notifications might make them feel disconnected or left out, especially if their friends are still getting them. They might check their phone more often at first, looking for something that isn't there. This is normal. The habit of reaching is still active even when the trigger is gone.

But after a few hours, something else starts to happen. You begin to choose when to check in rather than being summoned. You open an app because you want to see what's there, not because it told you to look. That small shift in agency matters. It's the difference between being interrupted and being intentional.

Some families notice they feel calmer. Meals last longer. Conversations don't get derailed as often. Bedtime routines go more smoothly because no one's phone is lighting up with one more thing. Kids sometimes report feeling less anxious, though they might not use that word. They just say things feel quieter or easier.

Other families don't notice much change, and that's useful information too. It might mean notifications weren't the main source of distraction in the first place. Or it might mean the habit of checking is so ingrained that it continues even without the prompt.

At the end of 48 hours, sit down with your child again and talk about what you both noticed. What felt different? What felt hard? Were there moments when you missed the alerts, or moments when you were glad they were off? Did anything surprise you?

Then decide together what happens next. You don't have to keep everything off forever. But you also don't have to turn everything back on. Maybe some notifications are genuinely helpful and others are just noise. Maybe your child realizes they don't actually care about likes on a post, but they do want to know when a friend messages them directly. Maybe you discover that turning off work email after a certain hour makes your evenings feel like your own again.

The point isn't to find the one right answer. It's to experience the difference between being pulled and being present, and then make a conscious choice about what you want your attention to feel like.

What this experiment reveals, more than anything, is how much of our device use isn't really chosen. It's triggered, prompted, engineered. When you remove the triggers, you get a clearer picture of what you actually want versus what you've been conditioned to respond to. And that clarity is something you can build on, one decision at a time.

Chapter 5

Synthetic Reality:

AI Content, Deepfakes, And Learning To See Clearly

,"When content can be generated instantly, discernment becomes a survival skill."

High Human Input — High AI Involvement

Human-Created	Mixed	AI-Generated

Human-Created Personally made	**Mixed** Partial assistance	**AI-Generated** Automated

More human choice — Less transparency

Human-Created	Assisted Creation	Advanced Assistance	Automated Generation
• Handcrafted by a person. • No AI tools involved.	• AI tools assist the process. • Supports, but doesn't dominate content.	• Significant AI guidance. • Heavily mixed content.	• Largely AI created. • Minimal human input.

More human choice — Less transparency

5.1 What AI-Generated Content Actually Is

Your teenager shows you a video of a celebrity saying something outrageous. A classmate shares a photo that looks like it came from last weekend's party, but something about it feels off. Your younger child asks if the voice message from their favorite cartoon character is real.

These moments are happening in homes everywhere, and they share something in common: the content might not have originated from a camera, a microphone, or a real event. It might have been created by artificial intelligence.

Understanding what that actually means doesn't require technical expertise. It does require updating our mental model of how images, videos, and audio can come into existence.

For most of human history, a photograph was evidence that someone stood in a place with a camera. A video meant someone filmed an event. A recording captured someone's actual voice. Even when we knew these things could be manipulated—cropped, edited, spliced together—there was still an original moment, an original recording, something real at the source.

AI-generated content works differently. It doesn't start with reality and modify it. It starts with patterns.

Think of it this way: if you asked a skilled artist to draw a cat, they wouldn't need a specific cat in front of them. They've seen enough cats to know what makes something look like a cat—the shape of the ears, the proportions of the face, the way fur catches light. They'd draw from their accumulated understanding of "catness."

AI systems do something similar, but with far more data and far more speed. They've been trained on millions of images, hours of audio, or vast amounts of text. They've learned the patterns—not by understanding cats or voices or writing, but by processing statistical relationships. When you ask the system to create an image of a cat, it generates pixels based on those patterns. No camera was involved. No cat existed.

The same principle applies to text, audio, and video. An AI can generate a paragraph that sounds like a news article because it has learned what news articles tend to look like in terms of structure, word choice, and rhythm. It can create a voice that sounds like your neighbor because it has learned the acoustic patterns of human speech. It can generate video of someone saying something they never said because it has learned how faces move when people talk.

This is different from Photoshop or traditional video editing in a fundamental way. Those tools manipulate something that existed. AI generation creates something from nothing—or more precisely, from pattern recognition applied to a prompt.

The word "deepfake" entered our vocabulary to describe AI-generated videos of real people, but the technology isn't limited to famous faces or obvious forgeries. It's increasingly accessible and increasingly convincing. A middle schooler with a laptop can generate an image of a classmate in a compromising situation. A scammer can clone a voice after hearing just a few seconds of audio. Someone with no video editing experience can create footage of an event that never happened.

None of this requires sophisticated equipment. The tools have become simple enough that the technical barrier is almost gone. What remains is intent—and the question of whether the viewer, listener, or reader can tell the difference.

Here's what makes this particularly challenging for families: our instincts about authenticity are based on older rules. We learned to be skeptical of obviously manipulated images—things that looked airbrushed or clumsily edited. We learned that if we heard someone's voice, that person said those words. We learned that video was harder to fake than still images.

Those instincts are now incomplete. AI-generated content can pass what researchers call the "eye test" and even the "ear test." It can look and sound completely real because it's built from the patterns of real things. The seams don't show in the way they used to.

Children and teenagers encounter this content in contexts where they're primed to trust it: messages from friends, posts in familiar online spaces, media shared by people they know. The content isn't labeled. There's no warning sign. And increasingly, there's no easy way to tell.

This doesn't mean every image or video is suspect. Most of what your family encounters on a given day is still traditionally captured and created. But the ground has shifted enough that we can no longer assume the old markers of authenticity hold.

The practical question isn't whether AI-generated content exists—it does, and it's proliferating. The question is what it means for your family to navigate a world where seeing or hearing something no longer settles the question of whether it's real.

5.2 The New Trust Problem

For a long time, the question people asked when they encountered information was simple: is this true? You could check a fact, trace a claim back to its source, verify a number. The internet made that harder in some ways and easier in others, but the basic architecture of truth-seeking stayed recognizable. If you wanted to know whether something was real, you looked for evidence.

That question is still being asked, but it's no longer the dominant one. Somewhere in the last decade, it got quieter. The question that rose up to replace it sounds almost the same, but it works completely differently: does this feel true?

This isn't about laziness or ignorance. It's about volume and velocity. Your teenager sees hundreds of claims in a day, maybe thousands. A headline about a new health risk. A video alleging something happened at a school across the country. A screenshot of a text conversation that might be real or might be fabricated. A statistic that sounds shocking. An image that looks wrong but not obviously fake. A celebrity quote that seems plausible.

There's no time to fact-check all of it. Even if there were time, many of these claims live in a gray zone where verification is difficult or impossible. The screenshot could be real. The anecdote could have happened. The photo could be from a different event but still mean something. So instead of asking "is this true," people start asking "does this fit with what I already think is true?" They pattern-match. They check it against their sense of how the world works, what kinds of things happen, who the good guys and bad guys are.

This shift didn't happen because people stopped caring about truth. It happened because the information environment became overwhelming, and feeling became a faster, more efficient sorting mechanism than verification. Your brain does this automatically. It has to. You can't live in a state of constant skepticism about every piece of information you encounter.

Here's where it gets tricky for families. Adults struggle with this, but they at least grew up in a world where verification was possible, even if it was slow. They remember what it felt like to wait for a newspaper correction, to look something up in an encyclopedia, to settle an argument by finding an authoritative source everyone agreed on. They have a mental model of how truth-seeking used to work, even if that model doesn't fit the current moment very well.

Kids don't have that model. They've never lived in a world where information moved slowly enough to be checked. They've never experienced a shared set of authoritative sources that most people trusted. For them, the question "does this feel true?" isn't a shortcut or a compromise. It's the primary method.

And the platforms know this. They're designed around it. The algorithm doesn't care whether something is true. It cares whether it's engaging, whether it provokes a reaction, whether it gets shared. A post that feels true to you gets amplified. A post that feels true to someone else, even if it contradicts the first one, gets amplified in their feed. Everyone ends up in a version of the internet that confirms what they already suspect.

This creates a strange problem for parents. You can't just teach your child to fact-check, though that's still worth doing. You also have to help them notice when they're operating in "does this feel true?" mode, because that mode is invisible to the person using it. It feels like clear thinking. It feels like common sense.

Imagine your child comes across a post claiming that a certain type of phone case causes cancer. The post has thousands of shares. There are comments from people saying they heard the same thing, that it makes sense, that big companies hide this kind of information all the time. The post might even include a photo of a scientific-looking chart. If your child's first instinct is to think "this feels true because companies do hide things," they're already operating in the emotional sorting mode. They're not asking whether the claim is verifiable. They're asking whether it fits their existing model of how the world works. And if that model includes a general distrust of corporations, the claim slots right in.

The hard part is that sometimes claims that feel true actually are true, and sometimes claims that feel false turn out to be accurate. The sorting mechanism isn't reliable, but it's not completely random either. This makes it difficult to talk about. You can't just tell your child "don't trust your instincts," because instincts matter. They're part of how humans navigate uncertainty.

What you can do is help them notice the difference between "this feels true" and "I've confirmed this is true." Those are two separate things, and conflating them is where the trouble starts. A claim can feel true and still be worth checking. An image can trigger an emotional response and still require context. A story can resonate with your sense of how the world works and still be incomplete or misleading.

This doesn't mean raising a child to be a skeptic about everything. Constant skepticism is exhausting and isolating. But it does mean helping them develop a kind of metacognition about their own reactions. When they encounter something that makes them angry, scared, or validated, that's a moment to pause. Not because the emotion is wrong, but because emotion is often the point. The thing they're seeing was likely designed to provoke that exact feeling.

The trust problem isn't going away. If anything, it's getting harder. Images and videos can be generated from scratch now, and they look real. Audio can be faked convincingly. Text can be written by machines that mimic human voices. The old verification methods are breaking down faster than new ones are being built.

But here's something worth holding onto: the skill of noticing when you're in "does this feel true?" mode is teachable. It's not about turning your child into a fact-checking machine. It's about helping them recognize the moment when their brain is taking a shortcut, and giving them the option to slow down if they choose to. Sometimes the shortcut is fine. Sometimes it's not. The key is knowing which one you're taking.

5.3 Deepfakes and Synthetic Media: The Practical Risks

A teenager receives a message from a classmate with an image attached. It appears to show her in a compromising situation she knows never happened. Within an hour, it's been shared across three group chats. By the time she tells a parent, dozens of people have seen it. The image is fake, generated by someone with a phone and a free app, but the damage is real and immediate.

This is not a distant problem or a theoretical concern. Synthetic media—images, videos, and audio created or manipulated using artificial intelligence—has become accessible enough that it no longer requires technical skill or expensive equipment. What once took a skilled professional hours in a video editing suite can now be done by a middle schooler in minutes.

The term "deepfake" originally referred to sophisticated video manipulations, often of public figures, that were difficult to detect. But the technology has evolved and spread. Now the bigger threat isn't elaborate video hoaxes. It's the quick, crude, but convincing-enough fake that gets made and shared before anyone can respond.

Consider how this plays out in practice. A student uses a classmate's social media photos to create a fake nude image. The creator might think it's a joke or a way to get attention. They might be angry or trying to humiliate someone. The motivation matters less than the speed. These images spread through social networks faster than any adult intervention can catch them. Even when they're taken down, screenshots persist. Even when everyone knows they're fake, the association remains.

The technology works by training artificial intelligence systems on enormous collections of images. Once trained, these systems can generate new images that follow the same patterns. If you feed the system photos of someone's face, it can place that face onto other bodies or into other contexts. The results aren't always perfect. Sometimes the lighting is wrong or the proportions are off. But they're often good enough to be believed at first glance, and first glances are what matter in a group chat or a social media feed.

Audio deepfakes follow a similar pattern. With a few minutes of someone's voice—pulled from a video they posted, a voice message they sent, or even a conversation recorded without their knowledge—software can generate new speech in that person's voice. Imagine a child receiving a panicked call that sounds exactly like their parent, asking them to take an urgent action. Imagine a parent receiving what sounds like their child's voice, saying they're in trouble and need money sent immediately. These scams are already happening.

The word "fake" almost undersells the problem because it suggests something obviously false. But synthetic media occupies an uncomfortable middle space. It's false in origin but can feel true in presentation. When we see a photo, we instinctively trust it as evidence of something that happened. When we hear a voice, we trust that the person is really speaking. Synthetic media exploits these instincts.

The harassment potential is obvious and immediate, particularly for young people still forming their sense of identity and reputation. But there's a second layer of harm that's less visible. When fake images and videos become common, trust itself starts to erode. A person who's been genuinely harmed might not be believed because "it could be fake." Someone caught doing something wrong might claim the evidence is synthetic. The technology creates plausible deniability where none should exist.

This uncertainty spreads beyond individual incidents. Consider a family where a teenager shows a parent a troubling video they saw online—maybe something political, maybe something about a local event, maybe something about a person they both know. The parent's first question might now have to be "is this real?" rather than "what do you think about this?" The conversation shifts from meaning to authenticity, from values to forensics. That shift has a cost.

Schools are already dealing with incidents where students create and share fake images of classmates or teachers. Some of these incidents are clearly malicious. Others start as jokes that spiral out of control. Either way, the institutions meant to protect young people are often slower than the technology spreading among them.

Law enforcement and legal systems are still catching up. Creating a fake nude image of a real person without consent is illegal in many places, but enforcement is inconsistent and complicated by questions of jurisdiction, proof, and age. A fake image created by one minor of another minor might be treated as a criminal matter, a school discipline matter, or both, depending on where it happens and who gets involved first.

The response from technology companies has been partial and reactive. Some platforms now label synthetic media when they can detect it. Some ban certain types of deepfake creation tools. But detection is always playing catch-up with generation. By the time one method is blocked or flagged, a new tool emerges. The apps that create these images often exist outside the major platforms, distributed through side channels or hosted in jurisdictions with minimal oversight.

Parents are left in an awkward position. They're expected to monitor and guide, but the technology moves faster than most people can learn about it. By the time a parent hears about a new app or technique, their children have likely already encountered it. This isn't a failure of parenting. It's a structural reality of how quickly synthetic media tools have spread.

The most practical risks are also the most intimate ones. These aren't foreign influence operations or elaborate hoaxes targeting strangers. They're personal attacks, impersonations, and manipulations happening within existing social networks. The person creating the fake image often knows the target. The people sharing it are often classmates or acquaintances. The harm is local and immediate.

There's also a learning curve that runs in the wrong direction. Adults who didn't grow up with this technology often assume they'll be able to spot a fake. Young people who have grown up seeing synthetic media are sometimes more skeptical, but they're also more desensitized. What shocks a parent might barely register with a teenager who's seen dozens of similar fakes. This gap in perception can make it harder to have conversations about what's serious and what's not.

The challenge isn't just teaching children not to create harmful content. It's helping them understand what's happening when they encounter it, how to respond when it involves them or someone they know, and how to think about truth and evidence in an environment where both can be manufactured. These are not simple lessons.

What makes synthetic media particularly difficult is that the harm often happens before the correction. A fake image gets shared, seen, and believed. Later, maybe hours or days later, someone proves it's fake. But the initial impact doesn't disappear. The embarrassment, the rumors, the changed perceptions—these persist even after the truth comes out. This is true for children targeted by peers, but it's also true for public figures, for anyone who becomes the subject of a viral fake.

The proportional response is somewhere between panic and dismissal. This isn't a reason to avoid all technology or assume everything online is false. But it is a reason to approach images, videos, and audio with more questions than we used to need. It's a reason to talk with children about what they're seeing and how they're interpreting it. It's a reason to recognize that visual evidence, once considered reliable, now requires verification.

The deeper shift is cultural and cognitive. We're moving from a world where most of what we saw was real by default to one where we have to actively consider whether something is real. That's a significant change in how we process information, and it's happening faster than our collective ability to adapt. Parents and children are navigating this shift together, neither of them with a clear map.

5.4 Persuasion at Scale

When a stranger tries to convince your child of something, you can usually see it coming. The tone shifts. The request feels off. There's a moment where instinct kicks in and warns you both that something isn't right.

But what happens when that stranger has already studied your child's fears, interests, and emotional patterns before the conversation even begins?

This is what separates older forms of deception from what AI-powered systems can do now. Traditional scams relied on generic scripts and luck. A phishing email might pretend to be from your bank, but it used the same wording for everyone. If you'd never had that particular account, the message fell flat. If you were having a calm day, you might pause and think it through. The scammer was fishing blind, hoping someone would bite.

AI changes the economics of that game entirely. Instead of one message sent to a million people, imagine a million messages, each one subtly different, each one adjusted in real time based on how the recipient responds. The system doesn't need to guess what will work. It tests, learns, and adapts faster than any human manipulator ever could.

Consider a teenager who mentions feeling anxious about an upcoming test in a group chat. An AI trained to recognize emotional vulnerability could generate a message that arrives hours later, framed as helpful advice, offering a service or product that promises to ease that exact worry. The tone matches how the teen already talks online. The timing feels organic. There's no obvious red flag, because the message wasn't written by a clumsy scammer following a template. It was tailored.

This isn't science fiction. Language models can already analyze tone, infer emotional states, and generate text that mirrors someone's communication style. When combined with data about what someone likes, what they've searched for, or what they've clicked on before, these systems can craft messages that feel personal and trustworthy even when they're neither.

The manipulation doesn't always look like a scam in the traditional sense. Sometimes it's subtler. A product recommendation that arrives exactly when your child feels insecure about their appearance. A political message designed to amplify anger about an issue they already care about, pushing them toward a more extreme version of their own beliefs. A suggestion to join a community that feels welcoming at first but gradually isolates them from other perspectives.

What makes this particularly difficult to guard against is that the messages often contain some truth. They're not entirely fabricated. They're just selectively framed, timed, and targeted to provoke a specific reaction. Your child might think they're discovering something on their own, when in reality they're being guided by a system optimized to hold their attention and shape their behavior.

The scale is what shifts everything. A human con artist can only manage so many targets. They get tired. They make mistakes. They can't personalize their approach for hundreds of people simultaneously. AI doesn't have those limitations. It can run the same persuasive tactics across thousands of interactions at once, learning from every response, refining its approach with each new data point.

This also means that the old advice—"just teach your kid to spot a scam"—doesn't cover the problem anymore. It's not enough to recognize bad grammar or suspicious links when the message is grammatically perfect and the link leads to a real website. It's not enough to distrust strangers when the voice on the other end sounds like a peer, or even like someone your child has been talking to for weeks.

The tools that make this possible aren't locked away in some shadowy corner of the internet. They're commercially available. Anyone with a basic understanding of how to prompt a language model can use it to craft persuasive messages. The barrier to entry has collapsed, which means the volume of sophisticated manipulation has grown exponentially.

None of this means your child is doomed to be deceived, but it does mean that resilience now requires something different than it used to. It's not just about skepticism toward obvious threats. It's about developing a deeper awareness of how persuasion works, how emotions can be leveraged, and how something can feel true and personal without actually being either.

What becomes important isn't teaching your child to distrust everything, which would be exhausting and isolating. It's helping them notice when their emotions are being played on, when urgency is being manufactured, or when they're being nudged toward a decision that benefits someone else more than it benefits them. That awareness doesn't come from a single conversation. It develops over time, through practice and reflection.

The persuasion your child encounters won't always announce itself. It will feel smooth, natural, maybe even helpful. But underneath, it's often optimized for one thing: keeping them engaged, clicking, buying, or believing whatever serves the system's goal. Understanding that changes how they move through digital spaces, not with paranoia, but with a clearer sense of when they're being led and when they're choosing for themselves.

5.5 How to Spot AI Without Becoming Paranoid

There's a moment many parents hit where they realize they can no longer trust their eyes. A video surfaces that looks completely real until someone points out the lighting is wrong. A voice message sounds exactly like a family member asking for help. A screenshot from a chat app seems legitimate until you notice the font is slightly off. The ground shifts, and suddenly everything feels suspect.

The instinct is to become a detective, scrutinizing every pixel and waveform. But that path leads to exhaustion and mistrust that seeps into everything. Kids pick up on it. They see you squinting at their phone, doubting their friends, questioning stories that might be completely true. The goal isn't to become an expert forensic analyst. It's to develop a quieter skill: noticing when something feels off, and knowing what to do with that feeling.

Start with the idea that you're looking for patterns, not proof. Imagine a parent sees a video of a local teacher supposedly saying something shocking at a school board meeting. The video could be real. It could also be edited, taken out of context, or generated entirely by AI. Instead of trying to determine authenticity frame by frame, the better question is: does this fit a pattern I recognize?

Real video has texture. People blink irregularly. They shift their weight. Their hair moves in ways that respond to actual physics. When someone talks, their words match the subtle movements of their throat and shoulders as they breathe. AI-generated video often smooths these details away. Faces might look waxy or oddly still between expressions. Hands tend to be the giveaway—fingers that blur together, extra joints, movements that don't quite track. Background elements sometimes warp or shimmer, especially near the edges of a person's body.

But here's the thing: these tells are getting harder to spot. The technology improves every few months. What looked obviously fake last year might look flawless now. So rather than memorizing a list of visual glitches, pay attention to context. Does the video appear suddenly with no clear source? Is it being shared with urgent language designed to make you react immediately? Does the situation it depicts seem designed to confirm something you already suspected or feared? Audio is trickier. Voice cloning has become sophisticated enough that a few seconds of someone's voice—pulled from a video they posted, a voicemail they left, a presentation they gave—can be used to generate something that sounds nearly identical. Imagine a scenario where a teenager gets a call that sounds exactly like their grandmother, claiming she's in trouble and needs money wired right away. The voice has the right accent, the right pacing, even the little throat-clear she does. But there's a flatness to the emotional tone. The urgent words don't quite match the cadence. Or the background is eerily silent, with none of the ambient noise you'd expect from a police station or hospital.

The instinct in that moment is panic. The better response is a simple verification step: hang up and call the person back at a number you already have. Use a different communication channel. If someone texts asking for help, call them. If someone calls, send a message through another app. AI can spoof one channel convincingly, but it's much harder to maintain the illusion across multiple platforms in real time. Screenshots present their own challenge. They're easy to fake—you don't need sophisticated AI, just basic editing tools. A fake text message, a fabricated social media post, or a manipulated email can look entirely real at a glance. The tells are often small: mismatched fonts, timestamps that don't align with the platform's current design, spacing that's slightly off. But the more useful question is the same one: does this fit a pattern of behavior you actually recognize? If a screenshot supposedly shows your child's friend saying something cruel, does that align with what you know about that kid? If it shows a teacher making an inappropriate comment, does it match their actual communication style?

Sometimes the content itself is the giveaway. AI-generated text has gotten remarkably good, but it still has tendencies. It loves certain phrases. It tends toward a kind of smooth, corporate reasonableness even when trying to sound casual. It struggles with specificity—real people mention brand names, recall weird details, go off on tangents. AI tends to stay on message. When you see a story or a post that feels oddly polished, that hits all the right emotional notes but lacks the messiness of real human communication, it's worth pausing.

Consider a situation where a teenager shows you a thread where someone is spreading rumors about them. The messages are mean, but they're also strangely articulate. Each insult is perfectly calibrated. There's none of the typos, the weird autocorrects, the fragments that characterize how teenagers actually text each other. It's possible the person writing is just unusually careful. It's also possible the messages were generated or heavily edited to maximize impact.

The deeper skill isn't learning to spot every fake. It's learning to sit with uncertainty. When something doesn't feel right, you don't need to prove it one way or another. You can simply treat it as unverified. You can say to your child, "I don't know if this is real, and we're not going to make decisions based on it until we know more." You can model the idea that it's okay not to have an immediate answer, that skepticism doesn't require certainty.

This is where the paranoia trap opens up. If you approach every image, every video, every message as potentially fake, you end up paralyzed. Worse, you teach your kids to trust nothing, which is its own kind of harm. The balance is in recognizing that most of what you encounter is probably real, but some of it isn't, and you don't always need to know which is which to respond appropriately.

Think of it like food safety. You don't inspect every bite you eat for contamination, but you do notice when something smells wrong or when a situation seems risky. You wouldn't eat gas station sushi from a warm cooler, but you don't treat every meal as a potential hazard. The same logic applies here. Most content is fine. Some isn't. You develop a sense for when something deserves extra scrutiny, and you build habits that protect you without consuming your attention.

One of those habits is checking sources. If a video makes a serious claim, see if it's being reported by organizations you recognize. If a screenshot shows something shocking, look for the original post. If an audio clip is going viral, try to find where it came from. You're not doing forensic analysis. You're just adding one more data point to your assessment.

Another habit is slowing down. AI-generated content often spreads through urgency. It wants you to react now, to share now, to feel something intensely before you have time to think. When you feel that pressure, it's a signal. Not necessarily that the content is fake, but that someone wants you to skip the verification step. Taking an extra minute, even an extra hour, usually reveals more than any technical analysis.

The goal is to remain engaged without becoming suspicious of everything. Your kids need to see you navigating this landscape with confidence and care, not fear and doubt. They need to learn that skepticism is a tool, not a worldview. That verification is smart, not cynical. That it's possible to stay connected and informed without believing everything or trusting nothing.

The technology will keep improving. The fakes will get better. The tells will get subtler. But the underlying approach remains steady: notice patterns, verify what matters, and accept that you won't always know for sure. That's not a failure of attention. It's the reality of living in a world where the tools for creating and manipulating content are now in everyone's hands, and the best response is neither blind trust nor constant vigilance, but something quieter and more sustainable in between.

5.6 Teaching Kids (and Adults) a Verification Habit

There's a moment that happens in most households when someone at the dinner table announces something they just learned online. Maybe it's about a food that's supposedly toxic, or a celebrity who died, or a new law that's coming. The announcement lands with certainty, and everyone reacts accordingly. Only later, if at all, does someone quietly discover it wasn't true.

We've all been on both sides of this. The internet rewards speed and emotion, not accuracy. And because so much of what we see is true, or true enough, we develop a habit of assuming. That assumption becomes the default, especially when something confirms what we already suspected or makes us feel outraged or validated.

The goal isn't to raise a generation of skeptics who trust nothing. It's to help kids develop a reflex that kicks in just before they believe or share something significant. A small pause. A quick check. Not paranoia, just care.

This reflex doesn't come naturally. It has to be modeled and practiced, the same way you'd teach a child to look both ways before crossing the street. The good news is that it's not complicated. The harder part is remembering to do it ourselves.

The Pause

The first step is simply noticing when something feels like it matters. Not every post requires scrutiny. But if your child is about to repeat something as fact, or forward it to a group chat, or argue with someone using it as evidence, that's a moment to pause.

You can frame this for younger kids as a gut-check question: "Does this feel big or important?" If yes, it's worth a second look. For older kids, the signal might be emotional. If they feel a strong reaction—anger, vindication, fear—that's often a sign the content is designed to provoke, not inform. Strong feelings don't mean something is false, but they do mean it's worth verifying.

Parents can model the pause by narrating their own process. "Hang on, this sounds too weird to be real. Let me check something." That simple act, spoken aloud, teaches more than any lecture.

The Check

Once you've paused, the next question is: where did this come from? For younger kids, this might mean helping them trace something back to its origin. If they saw a fact on social media, can they find the original article? If someone made a claim in a video, is there a source linked in the description? Often the answer is no. That absence itself is information.

Older kids can learn to skim past the headline and look at the URL, the publication name, or the account that posted it. A random string of words followed by ".com" isn't the same as a news outlet they recognize. A screenshot of a tweet isn't the same as the tweet itself. Teaching them to notice these differences doesn't require technical expertise. It just requires slowing down long enough to look.

One useful habit is the two-source rule. If something feels significant, see if at least two independent, credible sources are reporting it. Not two people sharing the same post, but two separate organizations or journalists who arrived at the information on their own. This doesn't guarantee truth, but it dramatically increases the odds. And if you can only find one source, or if every version traces back to the same dubious origin, that's a clue.

For images, reverse image search is surprisingly simple once you know it exists. On most phones, you can long-press an image and select "search image" or something similar. This reveals whether the photo is old, taken somewhere else, or digitally altered. Imagine a situation where a child sees a shocking image claiming to show a recent disaster. A quick reverse search might reveal the image is years old and unrelated. That one move can stop the spread of misinformation in its tracks.

The Confirm

Verification isn't about proving something wrong. It's about understanding it better. Sometimes the check reveals that yes, the thing is real and accurately described. Other times it turns out the claim is exaggerated, missing context, or entirely fabricated. Both outcomes are useful.

When you verify something with your child, walk them through what you're finding. "Okay, this article is from a site I've never heard of. Let me see if any major news organizations are covering this... no, nothing. That makes me wonder if it's real." Or: "This photo says it's from last week, but reverse search shows it's from three years ago. So the photo is real, but it's being used to support something misleading."

The process teaches them that truth isn't binary. Information exists on a spectrum from verified and trustworthy to misleading and false, with a lot of ambiguous territory in between. The habit isn't about always being right. It's about being less likely to be confidently wrong.

The Questions

Sometimes verification means asking better questions before accepting or sharing something. These questions can become second nature with practice:

Who benefits if I believe this? Who wants me angry or afraid? Does this fit too perfectly with what I already think? Am I being told what to feel? Is there more to the story than this snippet shows?

These aren't cynical questions. They're curious ones. They help kids recognize when they're being manipulated, not by some distant villain, but by the everyday incentives of online platforms that prioritize engagement over accuracy.

You can practice this during family conversations. When someone brings up something they saw online, instead of immediately debating whether it's true, ask together: what would we need to know to be sure? Where would we look? What are we assuming? These questions shift the dynamic from defensiveness to collaboration.

When Adults Need the Habit Too

Kids are often better at this than we are. They're digital natives, yes, but more importantly, they haven't yet spent decades building up the cognitive habits that make adults so susceptible to misinformation. We tend to trust sources that sound authoritative, share things that align with our worldview, and mistake familiarity for accuracy.

If you're teaching your child to verify, you have to be willing to do it yourself. That means catching yourself before forwarding something in the family group chat. It means admitting when you got something wrong. It means showing them that verification isn't about being smarter than other people—it's about being honest with yourself.

This isn't a skill you teach once and move on. It's a habit that has to be reinforced, the way you reinforce looking both ways or washing hands. Over time, it becomes automatic. Your child learns that information can be checked, that sources matter, that a moment of patience can prevent a cascade of confusion or harm.

What Verification Builds

Underneath this habit is something bigger than media literacy. It's the capacity to hold uncertainty without panic, to delay judgment without feeling powerless, to question authority without rejecting expertise. These are the qualities that help a person navigate a complex, often deceptive information landscape without becoming paralyzed or cynical.

When your child learns to pause, check, and confirm, they're not just avoiding misinformation. They're learning to think clearly under pressure. They're discovering that they have agency, that they don't have to accept everything at face value, and that the truth is usually worth the effort it takes to find.

That's the kind of resilience that lasts long after childhood, long after any specific technology becomes obsolete. It's a way of being in the world that protects without closing off, that questions without dismissing, that remains curious without being gullible. And it starts with a simple pause.

5.7 Reflection Exercise: "Would I Believe This If It Didn't Match My Feelings?"

There's a particular moment that shows up in families more often than we'd like to admit. A teenager shares something they saw online, something that confirms what they already suspected about a teacher, a classmate, or a news event. The parent nods along, maybe adds their own observation that fits the same pattern. It feels like connection, like you're both seeing the same truth. But neither of you stops to ask whether you'd give the same story the same weight if it pointed in a different direction.

This isn't about being gullible. It's about how human minds work. We're drawn to information that matches what we already think, and we're skeptical of information that doesn't. Psychologists call this confirmation bias, but you don't need the term to recognize the feeling. When something aligns with your existing beliefs, it lands softly. When it contradicts them, it feels like friction.

Kids do this constantly online. If they believe their school is unfair, they'll seize on examples of unfairness and scroll past counterexamples. If they think a celebrity is problematic, they'll collect receipts and ignore context. If they're convinced a supplement works or a conspiracy holds water, they'll find a thousand posts that say so. The algorithm knows this and serves them more of what they've already engaged with, which makes the pattern self-reinforcing.

The challenge for parents is that lecturing about bias almost never works. Telling your daughter she's only seeing one side of the story tends to make her double down. Explaining to your son that he's cherry-picking evidence just sounds like you don't trust his judgment. The conversation turns into a debate about who's right, and the actual skill you wanted to teach gets lost.

That's where this reflection exercise comes in. It's not about proving anyone wrong. It's about building a habit of noticing when your feelings are doing the sorting for you.

The core question is simple: "Would I believe this if it didn't match my feelings?" You can adjust the wording depending on what fits your family. Some parents ask, "Would I share this if it said the opposite?" Others go with, "If this story supported the other side, would I still think it was solid?" The exact phrasing matters less than the underlying move, which is stepping outside your immediate reaction and checking whether the information itself holds up or whether you're just glad it exists.

Imagine a parent and their teen scrolling through videos about a political issue they both care about. One video presents a claim that fits their shared perspective. Instead of simply agreeing, the parent pauses and says, "Okay, that sounds right to me. But let me ask myself, would I believe this if it was arguing the opposite? Would I check the

source, or would I just assume it's wrong because I don't like the conclusion?" That's the exercise. You're not interrogating your kid. You're modeling a kind of self-awareness that's rare and valuable.
This works better when you do it out loud about your own thinking, not theirs. If you catch yourself sharing an article because it confirms your frustration with something, you can say, "I'm realizing I didn't really look at where this came from. I just liked what it was saying." That admission does more than a lecture ever could. It shows that everyone falls into this pattern and that noticing it isn't a failure, it's a skill.
You can also make it playful. When your kid shows you something they're fired up about, you might ask, "What would someone who disagreed with this say? Not because they're wrong, but just, what's their version?" This isn't about false balance or pretending every perspective is equal. It's about recognizing that strong feelings are a signal to slow down, not speed up.
Sometimes the answer to the question is yes, you would still believe it. The information is solid, the source is credible, and your feelings just happen to align with reality. That's fine. The exercise isn't about doubting everything. It's about checking whether belief is coming from evidence or from comfort.
Other times, the answer is no, or at least "probably not." Maybe you'd be more skeptical if the claim didn't fit so neatly. Maybe you'd want a second source. Maybe you'd notice that the headline is doing a lot of emotional work that the actual facts don't support. That realization doesn't mean you have to reverse your opinion, but it does mean you're thinking clearly about how you got there.
The exercise also helps with a problem that's especially common in online spaces: the tendency to treat alignment as proof. If ten accounts are all saying the same thing and they all match your view, it feels like consensus. But online consensus often just means the algorithm showed you ten people in the same bubble. This reflection question interrupts that feeling and reminds you that agreement isn't the same as accuracy.
One thing to avoid is turning this into a gotcha. If your kid shares something and you immediately fire back with "Would you believe this if it said the opposite?" it's going to feel like an attack. The question works when it's genuine, when you're both curious about how you

arrived at your conclusions. It works when you ask it of yourself first and let them see what that looks like.

You also don't need to do this every time someone shares something online. If you turn every conversation into a critical thinking seminar, people stop talking to you. The goal is to plant the question so it becomes a reflex over time. Eventually, your kid might catch themselves mid-scroll and think, "Wait, would I care about this if it didn't make me feel right?" That's the shift you're after.

What makes this exercise different from standard media literacy advice is that it doesn't require your child to know how to evaluate sources, check credentials, or trace information back to its origin. Those are useful skills, but they take time and effort. This question is faster. It's a gut check that works in the moment, right when someone is about to believe, share, or defend a claim.

It's also surprisingly effective at defusing arguments. When two people are dug in on opposite sides of something, asking "Would we believe this if it said the opposite?" can create just enough space for both of them to step back. Not always, but often enough to matter.

The deeper lesson here isn't really about screens or information. It's about recognizing that your feelings are powerful guides, but not infallible ones. They tell you what matters to you. They don't always tell you what's true. Learning to hold both of those things at once—to care deeply and think clearly—is one of the most valuable skills a person can develop. It's also one of the hardest, because it requires a kind of humility that doesn't come naturally when you're certain you're right.

If you can help your kid start asking this question, even occasionally, you're giving them something that will serve them long after they've stopped caring what you think about their screen time.

5.8 Family Rules for Synthetic Media

Your daughter sends you a text: "Mom, I'm at Sarah's but her mom had to leave for an emergency. Can you Venmo her dad $40 for pizza and a movie? Here's his number." The message sounds exactly like her. But something nags at you.

You call her phone. She picks up on the second ring, confused. She's still at volleyball practice. She never texted you.

This isn't a distant threat anymore. Voice cloning tools can mimic someone after hearing just a few seconds of audio. Video manipulation software runs on ordinary laptops. A scammer doesn't need your child's phone. They need a voice sample from a TikTok video, a school presentation posted online, or a birthday message shared with relatives.

The technology that makes this possible isn't going away. Which means families need a new kind of agreement, one that accounts for a world where seeing and hearing are no longer enough.

Start with what you share. Most families haven't talked about voice and video the way they've talked about photos. But synthetic media tools learn from real recordings. A voice memo sent in a group chat, a video of your son singing at a recital, your daughter's voice in the background of a friend's Instagram story—all of these can be harvested and replicated.

This doesn't mean pulling your kids offline or stopping all recordings. It means being more selective. Before posting a video where your child speaks clearly for several seconds, ask whether it needs to be public. If you're sharing it with extended family, a private album or direct message works just as well. If your teen posts their own content, help them understand that their voice and likeness have value, and that limiting who has access to clear recordings isn't paranoia. It's reasonable.

Then establish a verification rule. This is the simplest and most important agreement a family can make: if someone asks for money, a pickup, a password, or any urgent help through a single channel—text, email, voice message, even a phone call from an unknown number—you verify through a second channel before acting.

That means if your child texts asking for something unusual, you call them. If someone claiming to be your child calls from a strange number, you hang up and call their known number. If your partner emails you a wire transfer request, you walk into the next room and ask them face to face.

This rule works because synthetic media usually arrives through one channel. The scammer has a cloned voice or a fake text, but they don't control your child's actual phone, your partner's actual email account, or the person standing in front of you. Requiring a second channel breaks the illusion.

Some families use a code word for emergencies, something only household members know. If your child calls from someone else's phone and says "pineapple" at the end of the sentence, you know it's really them. If the voice on the other end hesitates or deflects when you ask for the word, you know something's wrong.

Other families keep a short list of what never happens over text or voice message. Money requests. Changes to pickup plans. Sharing passwords or account access. Questions about Social Security numbers or bank details. These conversations always happen in person or through a secure video call where you can see the person's face in real time and ask a question only they would know the answer to.

The specifics matter less than the agreement itself. What matters is that everyone in the family knows the rules exist and understands why. A thirteen-year-old who knows that you'll always verify a request before sending money won't feel hurt when you call to check. A sixteen-year-old who understands that their voice can be cloned will think twice before posting a long rant on social media.

This kind of agreement also protects your kids from being manipulated. Imagine your son receives a panicked voice message that sounds exactly like you, saying you've been in an accident and need him to withdraw cash from an ATM and read the prepaid card numbers over the phone to cover the hospital bill. If he's been taught that you would never ask for that over a voice message, he'll pause. He'll call your actual number. He'll verify.

You can frame this conversation without frightening anyone. You're not telling your kids that danger is everywhere. You're telling them that technology has changed how trust works. Just like you taught them to look both ways before crossing the street, you're teaching them to verify before acting on a request that seems urgent or unusual.

It helps to practice. Role-play a scenario where someone pretends to be a family member asking for help. Walk through what your child would do. Let them see how easy it is to pause, verify, and confirm. Let them feel the difference between blind trust and informed caution.

As your children get older, this habit becomes second nature. They won't just use it at home. They'll use it when a college roommate's voice suddenly asks them to cover rent through a weird app. They'll use it when a message claims to be from their boss asking for sensitive information. They'll carry this reflex into a world where synthetic media is everywhere, and they'll be harder to fool because of it.

The real shift here isn't technological. It's cultural. For most of human history, hearing someone's voice or seeing their face meant they were present. That's no longer true. Adjusting to this reality doesn't require constant vigilance or suspicion. It just requires a small, consistent habit: verify through a second channel.

Your family's rules don't need to be complicated. They need to be clear, practiced, and respected. What gets shared publicly. What requires verification. What never happens over a single message. These agreements won't prevent every problem, but they'll make your household significantly harder to deceive.

And in a world where synthetic media is cheap, fast, and convincing, that kind of friction is worth creating.

Chapter 6

Digital Hygiene & Defense

"Hygiene isn't restriction. It's maintenance."

Hygiene isn't restriction. It's maintenance.

Reframes digital hygiene positively.

Healthy habits **stick best.**
Start young, stay **consistent.**

6.1 Privacy Basics for Modern Families

When your child opens a reading app to finish their homework, the app knows which page they're on, how long they spent there, and whether they tapped the dictionary for help with a word. When they watch a video about space, the platform logs what they watched, what they skipped, and what they clicked on next. None of this feels particularly intimate in the moment. It just feels like using technology. But these small interactions don't vanish when the screen goes dark. They're collected, stored, and often combined with hundreds or thousands of other fragments from across different apps and websites. Over time, they form something much more revealing than any single data point would suggest: a detailed profile of habits, interests, struggles, and patterns.

This is what we mean by a data trail. It's not dramatic. There's no single moment when something obviously goes wrong. Instead, it accumulates quietly in the background of ordinary digital life.

Consider a family that lets their eight-year-old use a tablet for school projects and weekend games. The child searches for information about dinosaurs, watches craft tutorial videos, plays a city-building game, and uses a reading app before bed. Each of those activities generates data that gets recorded somewhere. The search engine logs the queries. The video platform tracks viewing habits. The game collects play patterns and in-app behavior. The reading app monitors progress and comprehension.

Now imagine that same child doing these activities over the course of a year. The accumulated data starts to reveal preferences, learning challenges, emotional responses to certain content, times of day when focus drops off, and which kinds of rewards or prompts influence behavior. This information has commercial value. It can be used to target ads, shape recommendations, or train algorithms that predict what will keep someone engaged.

Most parents don't realize how much of this collection happens automatically. There's usually no notification when data gets logged. The privacy policies exist, but they're written in language designed for legal protection rather than genuine understanding. Agreeing to terms of service becomes a reflexive action, something you click through to get to the thing your child actually wants to use.

The companies collecting this information aren't necessarily doing anything illegal. In many cases, they're operating within the boundaries of current privacy laws, which are inconsistent and still catching up to how data actually moves through the digital economy. But legality and safety aren't the same thing. A practice can be perfectly legal and still create risks that parents would want to know about.

One of those risks involves the sheer number of places where data ends up. When your child uses an app, information doesn't just stay with that one company. It often gets shared with third-party advertisers, analytics firms, data brokers, and other entities whose names you've never heard. These companies might combine your child's data with information from other sources to build even more detailed profiles. Some of this data gets bought and sold. Some of it sits in databases that could be breached. Some of it gets used in ways the original company never intended.

Another risk is persistence. Digital records don't fade the way memories do. A interest your child had at age seven might still be connected to their online profile at age seventeen. Search queries, viewing habits, and behavioral patterns can follow someone across platforms and years. This creates a strange situation where a child's digital past becomes part of their present in ways that are difficult to predict or control.

There's also the question of who decides how this information gets used. As a parent, you might have strong feelings about whether your child's reading difficulties should be visible to advertisers, or whether their gaming habits should inform what products get marketed to them. But once the data exists in someone else's system, your ability to make those decisions diminishes significantly.

None of this means technology is inherently dangerous or that digital tools can't be genuinely useful. It means the exchange isn't as simple as it appears. What looks like a free app or a convenient service usually involves an implicit trade: access in exchange for information. That trade might be worth it in some cases. In others, it might not be. But it's hard to make that judgment without understanding what's actually being collected and where it goes.

The complexity of modern data collection also makes it difficult to rely on intuition alone. When you think about privacy, you might picture someone snooping through your emails or watching through a camera. Those kinds of violations are easy to recognize. But privacy concerns in the digital age are often more subtle. They're about patterns detected across thousands of interactions, inferences drawn from seemingly unrelated data points, and profiles built from behavior rather than explicit information.

Imagine a parent who notices their child has been researching symptoms of anxiety online. The child hasn't told anyone they're struggling, but the search history reveals something private. Now imagine that same search history being logged by a search engine, combined with data from other apps, and used to serve ads for mental health services or self-help products. The child hasn't chosen to share this information publicly, but it's no longer private. It exists in commercial databases, potentially accessible to anyone willing to pay for it.

This is where the conversation about privacy intersects with the conversation about childhood. Kids don't have the same ability as adults to anticipate long-term consequences or understand complex systems. They can't meaningfully evaluate a privacy policy or predict how their data might be used five years from now. They're navigating platforms designed by adults, optimized for engagement and profit, often without the developmental capacity to recognize what's at stake.

As a parent, your role isn't to become a privacy expert or to audit every technical detail of every app. That's not realistic. But it does help to understand the basic mechanics of what's happening so you can make decisions that align with your family's values. Privacy isn't an all-or-nothing proposition. It's a series of ongoing choices about what feels acceptable and what doesn't, given the specific circumstances of your child's life and your own comfort level with risk.

What makes this challenging is that the systems themselves are designed to be opaque. Companies benefit when data collection feels invisible and when privacy settings are difficult to navigate. Opting out often requires more effort than opting in. Default settings tend to prioritize data sharing rather than restriction. This isn't an accident. It's a design choice that reflects where the financial incentives lie.

Understanding privacy in this context means recognizing that every digital interaction creates a record, that these records accumulate into something larger, and that once information leaves your control it becomes very difficult to retrieve. It means knowing that your child's online activity isn't just about what happens in the moment. It's about what might persist, who might access it, and how it might be used in ways you never anticipated.

This doesn't have to lead to fear or avoidance. It can lead to thoughtfulness. When you understand how data trails form, you're better equipped to decide which activities are worth the trade-off and which ones might benefit from more careful boundaries. You can have conversations with your child about what information they're sharing and help them develop an instinct for recognizing when something feels too intrusive. You can check settings, read permissions, and ask questions about what companies plan to do with the information they collect.

The goal isn't perfect privacy. That's not achievable in a world where digital participation has become a basic part of daily life. The goal is informed privacy: knowing enough to make intentional choices rather than drifting into defaults that don't serve your family well.

6.2 Passwords, MFA, and Account Hardening

Your child's digital life sits behind a door. That door has a lock. And whether that lock is made of tissue paper or tempered steel determines whether a stranger, a hacker, or a predator can walk right in.

Most families don't realize how weak their locks actually are. A password like "Soccer2018!" feels solid. It looks complicated. It has a capital letter, a number, an exclamation point. But to anyone who knows what they're doing, it's trivial to crack. Not in a movie-hacking sense, with code scrolling down a screen. In a mundane, automated sense. Software can guess millions of passwords per second. It starts with the most common patterns. Names, years, favorite teams. It works through predictable substitutions. Then it runs through breached password lists harvested from old data leaks, the kind that happen quietly and constantly across the internet.

If your child uses the same password across multiple accounts, one breach anywhere means a breach everywhere. An old compromise from a gaming forum they joined three years ago becomes a key that unlocks their current email, their social media, even accounts you didn't know existed.

So the first step is simple and non-negotiable. Every account needs its own password. And every password needs to be long and random. Not clever, not memorable, not based on something meaningful. Random. Twelve characters minimum. A mix of letters, numbers, and symbols with no discernible pattern.

You might be thinking: how is anyone supposed to remember dozens of random passwords? The answer is that no one should try. Use a password manager. These are apps that generate strong passwords, store them securely, and autofill them when needed. Your child learns one master password to unlock the vault. Everything else is handled. This isn't a convenience feature. It's the only realistic way to maintain strong, unique passwords across the sprawling ecosystem of accounts most kids accumulate by middle school.

But even a strong password isn't enough on its own. Because passwords can still be stolen. Phishing emails that look exactly like a real service. Fake login pages that harvest credentials. Malware that logs keystrokes. Social engineering that tricks someone into handing over their password willingly.

This is where multi-factor authentication comes in. MFA requires a second proof of identity beyond the password. Usually, that means a code sent to a phone, generated by an app, or confirmed through a biometric like a fingerprint. The attacker might have the password, but without physical access to the second factor, they can't get in.

Imagine someone learns your teenager's Instagram password. Without MFA, they're in immediately. They can read private messages, impersonate your child, lock them out by changing the password, or worse. With MFA enabled, that stolen password is useless. The intruder hits a wall. Your child gets an alert that someone tried to log in. The compromise stops before it starts.

Not all MFA is equally strong. Text message codes are better than nothing, but they can be intercepted through SIM swapping, a technique where an attacker convinces a phone carrier to transfer a number to a new device. Authenticator apps are more secure. Hardware keys, small USB devices that provide physical proof of identity, are the gold standard. But for most families, an authenticator app strikes the right balance between security and usability.

Enabling MFA takes five minutes per account. It's almost always found in the security or privacy settings. Some platforms make it optional. Others don't offer it at all, which is itself a red flag. If a service your child uses doesn't support MFA, that's worth knowing. It changes how much you should trust that platform with sensitive information.

Account hardening goes a step further. It's about reducing the attack surface, the number of ways someone could slip in. That means turning off unused features, like email forwarding rules that could silently send copies of messages elsewhere. It means reviewing connected apps and services, the ones your child might have granted access to years ago and forgotten about. It means setting recovery options, like backup email addresses and security questions, that can't easily be guessed or socially engineered.

One commonly overlooked detail: the email address associated with an account is often the weakest link. If someone can access that email, they can reset passwords across every service tied to it. So the email account itself needs to be the most secured thing your child owns. Strong password. MFA. Recovery options that you, the parent, control or at least know about.

This isn't paranoia. It's proportion. The effort required to harden an account is minimal. The cost of not doing it can be devastating. A compromised account doesn't just mean lost access. It means exposure. Private conversations made public. Personal information sold or weaponized. Accounts used to target others, to spread malware, to sextort.

And here's the uncomfortable part: kids often don't see this coming. They think of their accounts as extensions of themselves, spaces they control. They don't think about the architecture underneath, the vulnerabilities, the ways those spaces can be invaded. They'll reuse a password because it's easy. They'll skip MFA because it's annoying. They'll click "yes" on a permission request without reading it because everyone does.

Your role isn't to manage every password or police every login. It's to make sure the foundation is sound. Set up the password manager together. Walk through enabling MFA on the accounts that matter most. Explain why it's not optional. Make it clear that this is part of being online, like locking the front door is part of living in a house. Once it's done, it fades into the background. Your child logs in the same way they always did, just with an extra second for the second factor. The password manager makes everything smoother, not harder. And beneath that smooth surface, the door to their digital life is no longer made of tissue paper. It's solid. It's locked. And it stays that way.

6.3 Identifying Scams and Social Engineering

A teenager receives a text that looks like it's from their bank, warning that their account has been compromised. A middle schooler gets a message on Discord from someone claiming to be a game developer offering free currency. Your college student opens an email that appears to be from their university's financial aid office, asking them to verify their login credentials immediately.

These aren't random occurrences. They're carefully designed attempts to exploit how our brains process urgent information, how we respond to authority, and how we behave when we think we're helping someone we care about.

Social engineering is the formal term for what's really just manipulation dressed up in digital clothing. Instead of picking a lock or breaking through a firewall, someone tricks a person into opening the door themselves. The goal might be stealing money, gaining access to accounts, collecting personal information, or sometimes just causing chaos. What makes it effective isn't sophisticated technology. It's the exploitation of normal human instincts: the desire to be helpful, the fear of missing out, the impulse to avoid trouble, the tendency to trust what looks official.

The mechanics are simpler than most people realize. Every social engineering attempt contains the same basic ingredients: a hook that grabs attention, a reason you need to act quickly, and a request that seems reasonable in the moment but isn't.

Consider a message that appears to be from a friend saying their phone was stolen and they need you to send money through a payment app. The hook is the friend's name and profile picture. The urgency is the emergency situation. The request sounds like exactly what a friend would need. But if you pause and think about it, several things don't add up. If their phone was stolen, how are they messaging you? Why would they ask for money through an app instead of calling from a borrowed phone? Why the pressure to act immediately?

That pause is what social engineering tries to eliminate.

Young people encounter these attempts constantly, but not always in forms that look like traditional scams. A message from someone posing as a recruiter for a modeling agency. An offer to make money quickly by receiving packages at home and reshipping them. A request to vote for someone in an online contest by clicking a link. A warning that their favorite streaming service is about to cancel their account. The variety is endless, but the pattern stays consistent.

The most effective approach isn't teaching kids to memorize a list of warning signs. It's helping them develop what you might call a skepticism reflex. Not cynicism, but a habit of asking a few basic questions before responding to any unexpected request: Do I know this person or organization? Was I expecting this message? Does this request make sense given what I know about how things actually work? What happens if I don't act immediately?

That last question matters more than it seems. Legitimate organizations don't demand instant responses to urgent problems that you didn't know existed five minutes ago. Banks don't send texts asking you to click links and verify your password. Schools don't email students requesting Social Security numbers out of the blue. Real emergencies involving people you know can be verified through a different channel.

One helpful mental model is the "second contact" rule. Before responding to any message that requests information, money, or account access, contact the person or organization through a different method you know is legitimate. If someone texts claiming to be your bank, hang up and call the number on the back of your card. If a message appears to be from a friend in trouble, call them directly instead of replying to the message. If an email looks like it's from a company you do business with, go to their website by typing the address yourself rather than clicking the link provided.

This approach works because social engineering relies on keeping you inside the world the scammer has created. The fake bank text includes a phone number to call. The spoofed email contains a link to a fake website that looks real. The impersonated friend keeps the conversation moving quickly toward the request. Stepping outside that constructed reality breaks the illusion.

The language in these messages often contains small tells, though relying on poor grammar and spelling isn't enough anymore. Some scams are extremely well crafted. More reliable indicators include mismatched details, unusual requests, and pressure tactics. A message from your bank that addresses you as "Dear Customer" instead of your name. An email from a company asking for information they already have. Any communication that threatens negative consequences if you don't act within a very short timeframe.

Younger kids face a different flavor of the same problem. They're more likely to encounter manipulation that promises them something exciting: free game items, exclusive access, the chance to meet a favorite influencer. The mechanism is the same. Click this link, share your login, download this app, give me this information. The

consequence might be a stolen account, but it could also be the beginning of a more serious situation where someone is gathering information about a child's habits, schedule, or vulnerabilities.
Teaching kids about this doesn't require scaring them about predators lurking behind every screen. It means helping them understand that people online aren't always who they claim to be, and that anyone offering something that seems too good to be true probably isn't offering it for generous reasons. If a stranger messages you claiming to work for your favorite game company and offering free currency, why would they choose you? How would they have found you? What do they get out of it?
The answers to those questions don't add up, and recognizing that gap is a skill that develops with practice.
One useful conversation to have periodically is about what kinds of information are worth protecting and why. Kids often don't realize that details that seem harmless in isolation can be pieced together to create a fuller picture. Your school name, your sports team schedule, your birthday, your pet's name, the street you live on. Individually, these feel like small facts. Combined, they become answers to security questions and context that makes an impersonation more convincing.
Parents can model this kind of thinking by talking through their own decision-making when they encounter something suspicious. "I got an email that looked like it was from our credit card company, but when I looked at the sender's actual email address, it was from a random Gmail account. That's how I knew it was fake." These small narrations help kids build a mental library of real examples without needing to experience the consequences themselves.
It's also worth acknowledging that everyone falls for something eventually. The attempts are getting more sophisticated, and some are genuinely difficult to distinguish from legitimate communications. If your kid does respond to a scam, the priority is damage control, not shame. Change passwords, contact the relevant companies, monitor accounts. The lesson isn't that they're gullible. It's that recognizing manipulation is harder than it looks, and now they know one more pattern to watch for.
The goal isn't creating paranoid kids who trust nothing and no one online. It's raising people who have a working understanding of incentives. When someone contacts you out of nowhere and asks you

to do something, what do they get out of it? If the answer isn't obvious or doesn't make sense, that's information worth paying attention to.

6.4 Device Hygiene Fundamentals

When we think about keeping our kids safe online, most of us picture conversations about stranger danger or social media drama. We're less likely to think about the Tuesday afternoon when a pop-up appeared promising a free Minecraft mod, or the Saturday morning when someone clicked "Remind Me Later" on a software update for the third week in a row.

But here's the thing: digital safety isn't just about who your child talks to or what apps they use. It's also about the invisible maintenance that keeps devices working the way they're supposed to. Think of it like a smoke detector. You can teach fire safety all day long, but if the batteries are dead, you've lost a critical layer of protection.

Device hygiene sounds technical, but it's really just a set of habits that keep the doors and windows of your digital house in good repair. When these basics are in place, a lot of common problems either don't happen or get caught before they turn into crises.

What Updates Actually Do

Software updates feel like interruptions. They pop up at inconvenient times, they take forever, and honestly, the device seemed fine five minutes ago. So it's tempting to keep clicking "Later" until the reminder stops appearing.

But updates aren't just about new features or a slightly different shade of blue in the menu bar. Most updates exist to patch security holes that didn't exist when the device was made. Imagine your house is well-built and sturdy, but over time, someone discovers that a particular brand of lock can be picked with a paperclip. The lock company sends you a free replacement. That's what a security update is.

When devices run outdated software, they're vulnerable to problems that have already been solved. It's like leaving that pickable lock on your door even after you know about the flaw. This matters especially for kids' devices, because children aren't usually thinking about security. They're thinking about the game, the video, the group chat. That's appropriate for their age. But it means the grownups need to handle the boring stuff that keeps the foundation solid.

The simplest approach is to turn on automatic updates wherever possible. Most phones, tablets, and computers have a setting that lets updates install overnight or during downtime. This removes the decision entirely. You're not saying yes or no each time. You're just making sure the device stays current the same way you make sure the smoke detector has batteries.

Browser Settings and Extensions

Your browser is the front door to the internet. If a child uses a browser to watch videos, play web games, do homework, or browse anything at all, then how that browser is configured matters.

Most browsers come with some built-in protections, but they aren't always turned on by default. Features like "block dangerous sites" or "warn before downloading" exist in settings menus that most of us never open. Taking ten minutes to review those settings and flip a few switches can prevent a surprising amount of trouble.

Some families also use browser extensions designed to add extra layers of protection. These might block ads that contain misleading links, warn about phishing sites, or prevent accidental downloads of suspicious files. Extensions aren't magic, but they can act like a second set of eyes, especially for younger kids who are still learning what to watch out for.

One thing to keep in mind: kids will sometimes disable protections if they find them annoying or if they interfere with something they want to do. This isn't necessarily sneaky. It's normal kid logic. If a warning keeps popping up and blocking a site their friend recommended, they might turn it off without realizing why it was there in the first place. So if you set up protections, it's worth mentioning why they exist and checking in occasionally to make sure they're still active.

Backups as a Safety Net

Most of the time, when we talk about digital safety, we're talking about preventing something bad from happening. But backups are different. Backups are for when something has already gone wrong.

Imagine your child's tablet stops working. Maybe it fell, maybe a software glitch corrupted the system, maybe ransomware locked it down. If there's no backup, everything on that device is gone. School projects, photos, messages that mattered. For a teenager, losing months of creative work or personal history can feel devastating.

Regular backups mean that the device itself is replaceable. The content isn't tied to one fragile piece of hardware. If something breaks or gets compromised, you restore from the backup and move on.

This doesn't have to be complicated. Most devices offer automatic cloud backups that run in the background. You can also set up periodic backups to an external hard drive if you prefer to keep things local. The key is making it automatic, because if it requires someone to remember, it probably won't happen consistently.

Backups also give you freedom to act decisively if a device is compromised. If something feels wrong and you're not sure whether malware is involved, you can wipe the device and start fresh without losing everything. That option is only available if you've been backing up all along.

Teaching Safe Download Habits

Adults often underestimate how confusing the internet can be for kids when it comes to downloads. A child searches for "free drawing app," and the results include actual apps, ads that look like download buttons, sites offering cracked software, and pop-ups claiming their device is infected and they need to click here to fix it.

To a kid who's still learning, these all look somewhat legitimate. The fake download button might be bigger and more colorful than the real one. The site offering the cracked version might have friendly language and cartoon mascots. There's no obvious villain twirling a mustache.

Safe download habits start with teaching kids to slow down and check a few things before clicking. Where is this file coming from? Is it from an official app store or a random website? Does the site look professional, or is it covered in flashing banners and misspelled words? If something is supposed to be paid but this site offers it free, why?

This isn't about making kids paranoid. It's about helping them develop a light sense of skepticism, the same way you'd teach them to think twice before opening the door to a stranger or handing out personal information to someone they just met.

For younger kids, the simplest rule is: only download from approved sources. If they want an app, it comes from the official app store. If they want a file for school, it comes from the school website or a link their teacher sent. If they're not sure, they ask. As kids get older and earn more independence, you can teach the reasoning behind those rules so they can start making their own judgment calls.

The Cumulative Effect

None of these practices is flashy. Turning on automatic updates doesn't feel like a parenting victory. Enabling a browser warning or setting up backups doesn't come with a clear before-and-after moment where you see the danger you avoided.

But over time, these small habits create a kind of structural resilience. They reduce the number of ways something can go seriously wrong. They make it easier to recover when problems do happen. And they quietly teach kids that maintaining things matters, that tools need care, that a little prevention is easier than a lot of repair.

You're not building a fortress. You're just making sure the basics are solid so you can focus your attention on the things that actually require conversation, judgment, and presence.

6.5 Incident Response 101

The hardest moment in digital parenting often isn't the years of slow decisions. It's the sudden discovery that something has already gone wrong.

Maybe you glance at a notification and see language that stops your heart. Maybe another parent calls with a screenshot. Maybe your child comes to you in tears. In that first minute, your body floods with adrenaline. Your hands might shake. You want to fix it immediately, to make it stop, to understand everything right now.

That impulse is natural and completely understandable. It's also the moment when many families accidentally make things harder for themselves.

When something troubling surfaces, most of us operate on instinct. We want to confront the child, demand explanations, or immediately delete whatever we've found. We might call the other kid's parents right away, or post about it in a community group to get advice. We're trying to protect our child and solve the problem. But in those first chaotic hours, well-meaning actions can destroy evidence, escalate conflict, or close off options we didn't know we had.

Think of it like discovering smoke in your house. Your first move isn't to start rearranging furniture. It's to figure out where the smoke is coming from and whether anyone needs to get out. Incident response works the same way. Before you act, you pause just long enough to preserve what matters and think one step ahead.

The first principle is simple: don't let anything disappear. If you've seen something on a device, take a screenshot or a photo of the screen with another phone. Capture the username, the timestamp, the context around the message. If your child or another adult is about to delete a conversation, ask them to wait. You don't need to explain why in that moment. You can say you want to understand what happened before anything changes. This isn't about surveillance or punishment. It's about making sure that if you need to involve a school counselor, a therapist, or in rare cases law enforcement, you'll have what they need to help.

Imagine a parent who discovers that their middle schooler has been receiving threatening messages from a classmate. The messages are cruel and specific. The parent's instinct is to march into the child's room, take the phone, and immediately text the other child's parent. But in doing so, they don't screenshot the messages first. The other parent, horrified, tells their child to delete everything. By the next morning, when the school wants to intervene, there's no record. Both kids give different versions of what happened. The situation becomes a bitter he-said-she-said, and nothing gets resolved.

It didn't have to go that way. A few photos of the screen would have changed everything.

The second principle is to separate immediate safety from everything else. If your child is in physical danger, or if someone is threatening imminent harm, you act on that right away. You don't wait to document or strategize. But most digital incidents aren't emergencies in that sense. They're serious, they're upsetting, but they unfold over hours or days, not seconds. That means you have a little time to think.

Consider a situation where a parent finds a group chat on their teen's phone that includes sexually explicit images. The parent's heart is racing. They want to know who sent them, whether their child participated, and how to make it stop. But before confronting anyone, they take a breath. They screenshot the relevant messages without opening or forwarding anything themselves. They don't interrogate their child in anger. They don't call other parents that night. They recognize that this situation might involve multiple families, school policy, and possibly legal concerns. So they document what they see, secure the device, and reach out to a school counselor or a family therapist the next day for guidance on next steps.

This doesn't mean staying silent or doing nothing. It means acting carefully instead of impulsively.

The third principle is to avoid contaminating your child's account of what happened. When you discover something upsetting, you naturally want to hear your child's side immediately. But if you come in with accusations, panic, or anger, you shape their response. They might shut down, agree with whatever you're saying to end the conversation, or scramble to tell you what they think you want to hear. Later, if a counselor or administrator needs to understand the sequence of events, your child's memory will be tangled up with your reaction.

Instead, let them tell you what happened in their own words first. You might say something like, "I saw something on your phone that worried me. I'm not angry, but I need to understand what's been going on. Can you walk me through it?" Then you listen. You don't interrupt with interpretations or blame. You let the story come out, even if it's messy or incomplete. There will be time to address lying, poor judgment, or broken rules. But in the first conversation, your goal is to understand, not to judge.

Imagine a parent who finds evidence that their child has been sending mean messages about a classmate. The parent is furious. They storm in and accuse the child of being a bully. The child, defensive and ashamed, denies everything or minimizes it. The parent doesn't learn that the messages started as retaliation for something the other kid did first, or that the child has been feeling anxious and out of control for weeks. Without that context, the parent can't actually help. They can punish, but they can't address what's underneath.

The fourth principle is to think about who else needs to know, and in what order. Some situations are private and stay within your family. Others involve other children, other parents, a school, or outside professionals. Before you make calls or send emails, map out who the key players are and what each one needs to know. If you're not sure, it's okay to consult with one trusted person first—a therapist, a school counselor, or a close friend with experience in these situations—before widening the circle.

Once information is shared, you can't take it back. A vague text to another parent might create alarm without giving them enough context to respond wisely. A detailed email to a teacher might trigger a formal process you weren't ready for. This doesn't mean you hide things that need to be addressed. It means you think through how and when to bring others in, so that everyone involved can act thoughtfully instead of reactively.

Consider a situation where a parent discovers their child has been part of a group chat that shared a stolen test. The parent's instinct is to immediately email the teacher and copy the principal. But they don't stop to think about whether their child was an active participant or just present in the chat, or whether other kids were involved. The email sets off a disciplinary process that sweeps up multiple students, some of whom had nothing to do with the cheating. The fallout is messy, trust is broken, and the parent realizes too late that a quieter conversation might have addressed the issue without so much collateral damage.

Not every incident requires a dramatic response. Sometimes the right move is smaller and quieter than you think.

The last principle is to take care of yourself while you're figuring this out. Discovering that something has gone wrong is genuinely distressing. You might feel like you failed, or like you don't know your child at all, or like the world is more dangerous than you realized. Those feelings are real, and they deserve space. But they shouldn't drive your decisions in the first few hours.

If you're too upset to think clearly, it's okay to step away for a bit. You can tell your child, "I need some time to process this. We'll talk more tomorrow." You can call a friend and vent without naming names. You can write out your feelings in a journal so they're not bouncing around your head. The goal is to give yourself enough room to calm down so that when you do act, you're responding to what's actually happening, not to your worst fears about what it might mean.

Incident response isn't about being perfect or having all the answers in the moment. It's about slowing down just enough to think before you move. It's about protecting the information you'll need, listening before you react, and making space for a thoughtful response instead of a panicked one. The situation won't feel less serious if you take a breath first. But you'll have a much better chance of handling it in a way that actually helps your child, your family, and everyone else involved.

6.6 Evidence Preservation for Parents

When something concerning happens online, your instinct might be to delete it immediately. A cruel message. A disturbing image. A conversation that crossed a line. The impulse to make it disappear is completely understandable. But that same instinct can eliminate the very documentation you might need later.

Evidence preservation sounds formal, like something that belongs in a courtroom drama. In reality, it's closer to keeping a receipt. You're creating a record that exists independently of the platform, the account, or the other person's decisions. Consider a situation where a classmate sends your child threatening messages over several weeks. If you report it to the school without documentation, you're relying on memory and interpretation. If you have screenshots with dates and times, you're showing a pattern.

The reason this matters goes beyond proving something happened. Digital content is remarkably unstable. An account can be deleted. A post can be edited or removed. A conversation thread can vanish if someone blocks your child or deactivates their profile. Platforms themselves change their policies, shut down features, or purge old content. What exists today might genuinely not exist tomorrow, and not because anyone is trying to hide evidence. The architecture just works that way.

Start with screenshots. They're imperfect, but they're also immediate and universally accessible. When you take a screenshot of something concerning, capture more than just the offensive content itself. Include the username, the timestamp, and any surrounding context that shows how the interaction unfolded. Imagine a parent discovering that their child received an explicit image. A screenshot of just the image tells part of the story. A screenshot that shows the sender's profile, the time it arrived, and the messages before and after tells a much fuller one.

Some platforms let you download entire conversation histories or account data. This is different from screenshotting individual moments. It's asking the platform to generate a file that contains everything: messages, photos, timestamps, metadata. Not every platform offers this, and the ones that do often bury the option in settings menus. But when it's available, it creates a more complete record than manually capturing screens one by one. The file you download is yours. It doesn't depend on the platform keeping that data accessible.

Timelines matter more than you might expect. Human memory compresses and rearranges events, especially stressful ones. Your child might remember a single bad incident, but the downloaded data shows it was actually a dozen smaller interactions spread across three months. Or the reverse: what felt like months of harassment turns out to have been an intense few days. Neither version is wrong, but the documented timeline gives you something concrete to reference when you're talking to a school administrator, a therapist, or law enforcement.

Context is easy to lose in screenshots. A single message might look cruel or confusing without knowing what came before it. Imagine a parent seeing their child's message that says "I wish you'd just disappear." Alarming on its own. Less so if the previous message was someone asking them to help hide a birthday surprise. You can't screenshot everything, but you can screenshot enough that the meaning is clear. If you're documenting a conversation, capture the flow, not just the flash points.

Storage is simpler than it sounds. Create a folder on your computer or in a cloud service you control. Name it something straightforward. Inside that folder, organize by date or by incident. Don't rely on your phone's camera roll. Phones get upgraded, photos get deleted during cleanups, cloud storage reaches capacity. Move the documentation somewhere deliberate and separate from your everyday digital life.

Reporting procedures vary wildly depending on who you're reporting to and why. Schools, platforms, and law enforcement all have different expectations. Some want printouts. Some want digital files. Some want you to use their specific reporting form and nothing else. You can't predict which format will be needed, so the practical approach is to have the documentation ready in multiple forms. Screenshots saved as image files. A simple written timeline in a document. Links to posts or profiles, knowing they might break. The goal isn't to become a digital archivist. It's to avoid the situation where someone asks for documentation and you realize you don't have it in the form they need.

There's an emotional dimension to keeping these records. Looking at hurtful content over and over takes a toll. You don't need to review the evidence constantly. You just need to know it exists and where to find it. Some parents find it helpful to write a brief summary note after creating the documentation, then put the actual files away. The note reminds them what they saved and why, without requiring them to revisit the screenshots themselves.

Not every concerning moment requires documentation. Your child and their friend having a snippy exchange probably doesn't need to be preserved for posterity. But when something feels like it might escalate, when you sense a pattern forming, when you have that tightness in your chest that says this is different, that's when documentation shifts from optional to important.
You're not trying to build a legal case from day one. You're creating a record that lets you see clearly what's actually happening, separate from fear or assumption. Sometimes that record will show you the situation is more serious than you thought. Sometimes it will show you it's less. Either way, you'll be working from fact rather than impression.

FIELD NOTE: The phone you don't take is the evidence you lose
I had a parent report a serious situation involving a trusted adult and their child. The parent suspected the child's phone contained messages that mattered, but they didn't take it immediately. They left the device with the child because they didn't want to "punish" them.

The child, scared, ashamed, and feeling like they were in trouble, started deleting messages. By the time the phone was turned over, key evidence was gone.

Then came the second surprise: the parents expected the phone back within 24 hours. But once a device becomes potential evidence, it may need to be retained for the case, even if it was surrendered voluntarily, because the court can require it for chain of custody. That's not law enforcement being difficult; that's the justice system being strict about what counts as reliable evidence.

If you suspect a device contains evidence: take it, preserve it, and replace it temporarily with a basic phone for safety and communication. Your child's comfort matters, but so does protecting them with proof.

Chapter 7

Family Systems & Digital Mindset

"Systems reduce conflict by removing guesswork."

7.1 Building Your Family Operating System

"Without a system, every boundary becomes a debate."

Most families stumble into their screen routines. A device gets handed over during a stressful moment. A rule gets made on the spot. A boundary shifts because enforcing it felt too hard that day. Before long, you have a collection of inconsistent decisions that nobody quite remembers agreeing to, and everyone interprets differently.

This isn't a failure of parenting. It's what happens when technology moves faster than our ability to think through what we actually want.

A family operating system is different. It's not a rigid set of rules carved in stone. It's a shared understanding of how your household works—what you value, what you're protecting, and how you'll handle the inevitable friction that comes with screens. Think of it less like a contract and more like a language everyone in the family speaks.

The reason this matters is simple: without a shared framework, every screen decision becomes a negotiation from scratch. Your child asks for a new app, and you're weighing risk, fairness, precedent, and your own exhaustion all at once. Your partner makes a call you don't agree with, and suddenly you're arguing about TikTok when you're really arguing about consistency. When there's no operating system, every moment is improvisational, and that's exhausting for everyone.

Start by identifying what you're actually trying to protect. Not in abstract terms like "safety" or "balance," but in specific, lived terms. Are you trying to preserve uninterrupted family dinners? Protect sleep? Make sure your child has time for face-to-face friendships? Prevent exposure to content that's too mature, too cruel, or too addictive? These priorities will be different for every family, and they'll shift as your children grow. The point isn't to find the perfect answer. It's to know your answer.

Once you're clear on what matters, you can start building principles instead of rules. A principle might be: "We don't let screens replace human connection when someone's upset." That principle could mean no phones during arguments, no texting apologies for big conflicts, and no scrolling as a substitute for talking through a hard day. It's flexible enough to adapt as situations change, but firm enough to guide decisions in the moment.

Imagine a parent noticing their teenager reaching for their phone every time the conversation gets uncomfortable. Without a shared principle, that moment might pass unnoticed, or it might spark a reactive lecture about phone addiction. With an operating system in place, the parent can name what's happening in terms the family already understands. "Hey, I think we're doing the thing we talked about—using the phone to avoid the hard part." The teenager might still resist, but the conversation is grounded in something you've agreed on together, not something being imposed in the heat of the moment.

This is also where you address the reality that parents and caregivers need to be on the same page. If one adult allows YouTube at breakfast and another doesn't, your child isn't learning about boundaries. They're learning about leverage. It helps to have a handful of non-negotiables that everyone enforces the same way, and then some areas where you allow flexibility depending on who's in charge that day. The key is making those distinctions explicit, so your child isn't constantly testing to see which adult will give in.

Consider a situation where two parents disagree about whether their child should have a phone in their bedroom at night. Instead of each lobbying for their position, they might start by asking what they're each worried about. One might be worried about sleep disruption. The other might be worried about their child feeling untrusted or isolated. Once those concerns are on the table, the conversation shifts. Maybe the phone stays out of the bedroom, but they agree to check in more intentionally before bed. Maybe the phone stays in the room but goes into a lockbox at a set time. The specifics matter less than the fact that both parents understand the reasoning and can explain it the same way.

Your operating system also needs to account for the fact that your child is growing. A framework that works for a seven-year-old won't work for a fourteen-year-old. Build in moments to revisit and adjust. Maybe that's once a year, or at the start of each school year, or whenever your child asks for a significant new privilege. These conversations shouldn't feel like performance reviews. They're check-ins. What's working? What's not? What's changed?

One of the most useful things an operating system can do is clarify what happens when things go wrong. Not if—when. Your child will break a rule, hide something, or push a boundary in a way that surprises you. If you haven't talked through consequences in advance, you'll be making decisions in anger, fear, or confusion. That rarely goes well.

Consequences should be proportional, predictable, and connected to the behavior. If your child stays up past the agreed time on their device, the consequence might be losing access to that device in the evening for a few days. If they're caught on an app they promised not to use, you might revisit access to that category of apps altogether. The goal isn't punishment for its own sake. It's helping your child understand that agreements have weight.

But consequences alone don't build the operating system. You also need space for repair. When your child messes up, can they come to you and admit it? Or have you made the cost of honesty so high that lying becomes the safer bet? If your system doesn't allow for mistakes without catastrophic fallout, you'll lose access to the truth. And without the truth, you can't help.

There's one more piece that often gets overlooked: your own screen behavior. Children are extraordinarily good at spotting hypocrisy. If the family principle is "no phones during meals," but you're checking work emails between bites, that principle becomes a rule for kids, not a value for the household. If you're scrolling on your phone while half-listening to your child talk about their day, you're teaching them that screens are more important than presence, no matter what you say out loud.

This doesn't mean you have to be perfect. It means you have to be consistent enough that your child can trust the framework. And when you slip—because you will—you can name it. "I was on my phone too much this weekend. I noticed it, and I'm going to do better this week." That kind of honesty strengthens the system. It shows your child that the operating system applies to everyone, and that noticing when you fall short is part of how it works.

What you're building here isn't control. It's clarity. You're creating a set of shared expectations that reduces friction, increases trust, and gives everyone in the family a way to talk about screens without starting from scratch every time. It won't eliminate conflict, but it will give you a foundation to stand on when conflict comes. And it will come.

FIELD NOTE: Phones in steps—not all at once

In my house, a phone isn't a trophy you hand over because of a birthday or peer pressure. It's a tool—and tools come with training.

My daughter started with a basic phone with GPS and limited features. The rule was simple: show responsibility before earning more capability. That meant no use at school, keeping it charged, carrying it in public so we can reach her, and responding when we check in. Those weren't "gotcha rules." They were safety habits.

Now that she's getting older, we're considering a smartphone—but not because "all her friends have one." If we do it, we'll start with a lower-end device first. Not as punishment, but as a way to learn care, boundaries, and routine without turning an expensive phone into a constant negotiation. As trust builds, the phone can "level up."

This approach lowers risk and lowers arguments. The phone becomes part of a system: responsibilities first, privileges next, upgrades last.

7.2 Modeling: The Silent Curriculum

Your daughter asks to watch another episode. You say no, screen time is over. Five minutes later, you're scrolling through your phone while she does her homework at the kitchen table. She glances over. She doesn't say anything. She doesn't need to.

Children are scientists of the adults around them. They notice everything. They watch how you respond to a notification buzz, whether you bring your phone to the dinner table, how long it takes you to look up from a screen when they speak to you. They absorb these patterns the way they once absorbed language itself: not through instruction, but through observation and imitation.

This creates an uncomfortable reality. The rules you set about screens matter less than what you do with your own. A teenager who hears "put your phone away" from a parent who checks email during every conversation learns something, but it isn't about healthy boundaries. They learn that the rules are for children, that adulthood means doing whatever you want, or perhaps that the phone really is more important than the person in front of you.

The research on this is consistent and a little humbling. When parents describe their own phone use as problematic, their children's screen habits tend to follow the same pattern. When parents are frequently distracted by devices during family time, children report feeling less important, less heard. The message isn't delivered through words. It arrives through absence, through fractured attention, through the micro-moment when a parent's eyes leave a child's face and drop to a screen.

What makes this difficult is that most adults don't use devices the way children do. You might pick up your phone twenty times in an hour, but each glance feels purposeful to you. Checking a work message, responding to a text from your partner, looking up the name of that actor in the movie you watched last week. To your child, though, these distinctions blur together. What they see is a pattern: the phone calls you back, over and over. You answer.

Consider the parent who limits video games to weekends but spends weeknight evenings scrolling social media on the couch. The parent has reasons. Social media feels passive, a way to unwind. Video games seem more intense, more consuming. But to a child observing this, the taxonomy doesn't matter. What matters is that the adult gets to zone out with a screen and the child doesn't. The parent might explain the difference, might argue that adult judgment is more developed or that the content is different. The child hears this. They also notice that when the parent is scrolling, they're just as gone, just as unreachable.

None of this means you need to perform perfect digital habits or renounce screens entirely to have credibility. Children are surprisingly sophisticated observers. They can understand nuance. What they struggle with is hypocrisy, particularly the unacknowledged kind. When a parent says "I know I'm on my phone too much" or "I'm working on this too," it shifts the dynamic. The child isn't watching a hypocrite. They're watching a human being who is also navigating something challenging.

There's a particular moment that reveals the power of modeling more clearly than any study could. Imagine a family where the parent decides to put their phone in another room during dinner. No announcement, no new rule for the kids, just a quiet change in habit. The first night, maybe nobody notices. The second night, a teenager makes a sarcastic comment. By the end of the week, the tone at the table has shifted slightly. Conversations last longer. Someone tells a story that wouldn't have emerged in a shorter exchange. Eventually, without being asked, one of the kids starts leaving their phone behind too.

This isn't guaranteed to happen, and it isn't a strategy. It's just an example of what becomes possible when the modeling shifts. Children are so good at reading adults that they respond to changes in behavior before the behavior is even discussed. They're wired to learn by watching. That wiring doesn't turn off just because the lesson is about screens.

The hardest part might be accepting that you can't narrate your way out of this. You can't explain to a seven-year-old that your phone use is different because you're managing a household and they're just watching videos. You can't convince a teenager that the rules are fair when your own habits suggest otherwise. What you can do is let your actions teach what your words cannot. That's what modeling is: the curriculum that runs underneath every conversation, every rule, every boundary you try to set.

Your children are already learning from you. The question is what they're learning.

7.3 The Weekly Tech Check-In Ritual

Most families talk about screens only when something has already gone wrong. A child refuses to put the phone down at dinner. A parent discovers a troubling conversation. A teacher sends an email about declining focus. By the time the conversation happens, everyone is defensive, and the discussion becomes about blame instead of understanding.

A weekly tech check-in works differently. It creates a predictable, low-stakes moment when everyone expects to talk about how technology is actually functioning in your home. Not whether it's good or bad in the abstract, but whether it's working for your family right now.

The ritual takes about ten minutes. You pick the same time each week—Sunday after breakfast, Saturday before dinner, whatever fits your rhythm. The consistency matters more than the day. When it becomes expected, it stops feeling like an interrogation.

Start by asking your child to show you their most-used apps from the past week. Most phones surface this information automatically in screen time settings. You're not hunting for evidence. You're just looking together at what actually happened, the way you might review a week's worth of spending or look at a calendar to see where the time went.

Sometimes a child will be surprised by what shows up. They thought they spent an hour on a game, but it was closer to six. They remember texting friends but forgot about the two hours scrolling a feed. The data itself becomes the teacher. You don't need to lecture.

Ask how they felt during the week. Not "do you think you're on your phone too much," which invites a scripted answer, but something more specific. Did anything online make them feel worse this week? Was there a moment when they wanted to stop using something but found it hard? Did they notice their mood shift after certain apps? These questions sound simple, but they teach a child to pay attention to their own experience. Most of us, adults included, don't naturally connect our emotional state to our digital consumption. We feel irritable or scattered and don't trace it back to the hour we just spent comparing our lives to curated highlight reels. Building that awareness early is more valuable than any specific rule.

Next, look at settings together. Privacy controls change. Apps update and reset permissions. A messaging app that didn't have location tracking last month might have it now. This part of the check-in isn't about control. It's about teaching your child that these systems are designed to take more than they give back, and that paying attention is part of using them responsibly.

You might also use this time to add or remove an app. Maybe your child wants to try something new, and you agree to a two-week experiment. Maybe an app that seemed fine has become a problem, and you both decide to delete it. Framing these decisions as experiments rather than punishments changes the emotional texture of the conversation.

One underrated aspect of this ritual is that it forces parents to stay current. If you only check in when there's a crisis, you fall behind. You don't know what apps are popular, how interfaces have changed, or what your child's actual digital life looks like. Ten minutes a week keeps you oriented. You won't understand everything, but you'll understand enough.

The check-in also catches problems while they're still small. A child mentions feeling anxious after using a particular app. You notice they're getting notifications at two in the morning. They admit a group chat has turned mean. These are things you can address before they calcify into bigger issues.

Some parents worry this ritual will feel invasive, especially with older children. But when it starts early and remains consistent, it doesn't feel like surveillance. It feels like maintenance. You're not investigating. You're collaborating on something you both have to manage.

The tone you set matters. If you approach the check-in as a detective looking for violations, your child will hide things. If you approach it with genuine curiosity—what's working, what's not, what did we learn—they'll usually tell you the truth. Not always, but more often than you'd expect.

This isn't a conversation about whether technology is good or bad. It's a conversation about whether this specific technology, in this specific week, served your child well. Sometimes the answer is yes. Sometimes it's no. Sometimes it's complicated. That's fine. You're building a habit of reflection, not trying to reach a verdict.

Over time, something shifts. Your child starts noticing patterns without prompting. They'll tell you an app is making them feel bad before you ask. They'll suggest changes to their own settings. They'll start thinking of their devices as tools they manage rather than forces that manage them.

The ritual becomes a mirror. It doesn't tell your family what to do, but it shows you what's actually happening. And once you can see clearly, the decisions tend to make themselves.

7.4 The Graduated Trust Ladder

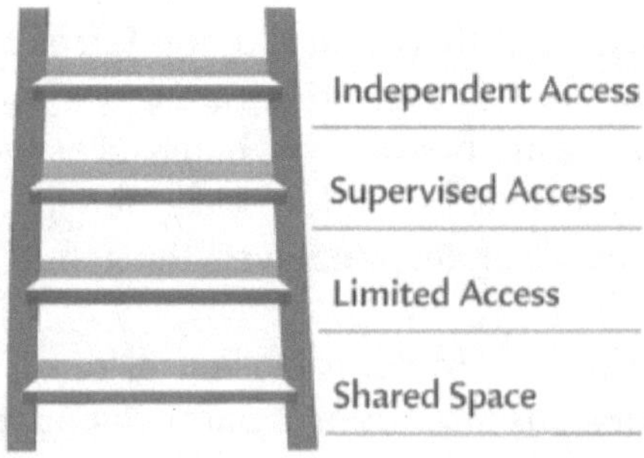

When your daughter asks for her own social media account, the question isn't really about age. It's about whether she's ready to handle something you can't always see or control. When your son wants to take his phone into his bedroom at night, you're not just deciding on a rule. You're trying to figure out if he's developed the judgment to stop scrolling when he should be sleeping.

These decisions feel hard because they are hard. You're being asked to predict how your child will behave when no one's watching. And the standard parenting approach doesn't help much here. Either you say yes and hope for the best, or you say no and delay the question. Neither one teaches your child how to manage freedom responsibly.

The graduated trust ladder works differently. Instead of granting independence all at once, you build it in stages. Each stage requires your child to demonstrate specific skills before moving forward. They prove they can handle a little freedom, and that earns them a little more.

Imagine a family where the eleven-year-old wants to start texting friends. Instead of simply handing over a phone, the parent starts with something smaller. The child gets a basic device that can only contact a short list of family members. For two months, the expectation is simple: respond to parent messages within an hour during waking hours, and plug the phone into the kitchen charger every night by eight. That's it. No complicated rules. Just demonstrate that you can manage a communication device reliably.

If the child does that well, the next step might be adding a couple of friends to the contact list. The phone still lives in a shared space at night, but now there's a bit more social access. A few months later, if things continue to go smoothly, maybe a basic messaging app gets added. Each expansion happens only after the previous level has been handled responsibly.

The ladder isn't about earning rewards for good behavior in general. It's not a gold star chart. The child has to show competence in the specific domain where they're asking for more freedom. A kid who wants to use YouTube unsupervised needs to first demonstrate good judgment with guided YouTube use. A teenager who wants a later curfew needs to show up on time consistently with the current one.

This approach changes the conversation between parent and child. Instead of arguing about whether they're "old enough" or whether you're being "too strict," you're both looking at the same evidence. Did they do what they said they'd do? Did they handle the responsibility well? If yes, they move up. If not, you stay where you are until they can.

It also gives you a way to respond when things go wrong without dismantling everything. Suppose your child gets their first smartphone and within a week you discover they've been staying up until midnight watching videos. In an all-or-nothing system, you might take the phone away entirely, which feels like a punishment and often leads to resentment. With a trust ladder, you simply move back down a rung. The phone goes back to the kitchen at night, and you revisit the conversation in a month. It's not about blame. It's about recognizing that they weren't quite ready for that level of independence yet.

The structure also makes it easier to have conversations about honesty. When your child knows that lying will reset their progress, the incentive shifts. They're less likely to hide a mistake if they understand that coming forward keeps them on the ladder, while covering up moves them backward. You might tell them explicitly: if you tell me you went over your screen time limit before I find out on my own, we'll talk about it and adjust. If I discover it because I checked and you didn't mention it, we move down a step.

Some parents worry this approach is too transactional, that it reduces trust to a series of checkboxes. But trust in real life is transactional. Adults lose access to things when they misuse them. You don't get to keep your driver's license if you repeatedly drive drunk. Your bank doesn't extend you more credit if you default on payments. Teaching children that freedom expands with demonstrated responsibility isn't cold. It's realistic.

The ladder also protects you from one of the more corrosive dynamics in family technology use: the constant negotiation. When every request for more access becomes a debate, everyone gets exhausted. Your child feels like you're arbitrary and controlling. You feel like you're under siege. The ladder removes most of that friction. The terms are clear. Your child knows what they need to show you to move forward. You know what you're watching for.

What makes the system work is that you have to genuinely give more freedom when your child earns it. If you keep finding reasons to stay on the same rung even when they've met the criteria, the ladder collapses. They'll stop trying. So when your child does demonstrate good judgment for a sustained period, you follow through. You let them have the later curfew, the phone in their room, the unsupervised access. That's the trade. They show you they can handle it, and you give them the chance to prove it at the next level.

Over time, the ladder becomes less visible. Your teenager isn't thinking "I'm on rung seven of the family trust system." They're simply living with the understanding that how they use their freedom today shapes what they'll have access to tomorrow. Which is, more or less, how adult life works anyway.

7.5 Planning for "When, Not If"

There's a moment most parents dread. Your child comes to you looking shaken, or you discover something on a device that wasn't supposed to be there, or a notification pops up that makes your stomach drop. In that instant, everything you thought you knew about your family's digital life suddenly feels fragile.

The mistake parents often make is treating these moments as failures of the system. We set up filters, establish rules, have conversations, and then feel blindsided when something slips through anyway. But the truth is simpler and less catastrophic than it feels: mistakes are how children learn to navigate complexity. The question isn't whether your child will encounter something unexpected online. It's whether they'll tell you when they do.

Imagine a situation where a ten-year-old searches for a music video and ends up three clicks away watching something violent or sexual that wasn't flagged by any filter. Or consider a middle schooler who accepts a friend request from someone whose profile looked normal but whose messages quickly become uncomfortable. These scenarios don't require catastrophic parenting failures. They just require the internet to work exactly as it's designed: fast, associative, and indifferent to context.

The instinct in these moments is often to treat them as emergencies requiring immediate correction. Parents confiscate devices, interrogate kids about how this could have happened, or launch into lectures about responsibility. All of this is understandable. But from the child's perspective, what they learn in that moment isn't about safety. It's about whether honesty is safe.

A "when, not if" mindset starts before anything goes wrong. It means acknowledging out loud, with your kids, that the digital world is designed to pull their attention in unexpected directions. You might say something like, "I know you're going to see things online that surprise you or make you uncomfortable. That's not your fault. It's how these platforms work. What matters is that you can tell me when it happens."

This isn't permission. It's accuracy. And children can feel the difference between a parent who's pretending the internet is manageable and one who's being honest about what they're up against.

When something does happen, the first few seconds of your response matter more than anything else. If your child comes to you and says they saw something disturbing, or someone sent them a weird message, your initial reaction will determine whether they come to you next time. This doesn't mean pretending you're not concerned. It means leading with curiosity before correction.

You might start with, "I'm glad you told me. That sounds confusing," or "Thank you for bringing this to me. Let's figure this out together." The goal is to keep the conversation open long enough to understand what actually happened, not just what you fear might have happened.

Imagine a parent whose teenager admits they created a second social media account their parents didn't know about. The immediate response might be anger or a sense of betrayal. But if the parent can pause and ask, "What were you hoping that account would give you?" they might learn their child felt suffocated by having extended family comment on everything they posted. That's a different conversation than one about deception. Both matter, but only one of them starts with understanding.

Scripts can help when your brain is moving faster than your words. You don't need to memorize them, but having a few anchors can keep you steady. When your child shows you something upsetting: "That must have been hard to see. Do you want to talk about it now, or do you need a little time?" When they admit to breaking a rule: "I'm not happy about this, but I'm glad you're telling me. Let's talk about what happened and what we do next." When you discover something they didn't tell you: "I found something on your device that worries me. I'm not going to yell, but I do need to understand what's going on."

These aren't magic phrases. They're simply ways to buy yourself a few seconds to think instead of react.

There's a practical element to this, too. If you wait until something goes wrong to figure out your household's response plan, you'll be making decisions in the worst possible state: flooded with adrenaline, imagining worst-case scenarios, and trying to look calm in front of your child. It's worth sitting down with a co-parent or a friend and talking through a few hypotheticals ahead of time. What would you do if your child told you someone online asked them to keep a secret? What if they admitted to lying about their age on a platform? What if they sent a message they now regret?

You don't need a perfect answer for every scenario. You just need to have thought about it once when you weren't panicking.

One thing that helps is remembering that most digital mistakes are recoverable. A child who sees something disturbing can be supported through it. A teenager who shares something they shouldn't can learn why privacy matters. Even serious situations, ones that genuinely require intervention, are almost always more manageable when a child feels their parent is on their side rather than just enforcing consequences.

This doesn't mean consequences don't matter. Sometimes taking a device away for a while, or pausing access to a specific app, is the right move. But even those decisions can be framed as protective rather than punitive. "I'm hitting pause on this for now because I think you need some space to reset, and I need time to figure out how to help you stay safer" is different than "You've lost my trust."

The hardest part of this approach is that it requires you to separate your child's behavior from your own fear. When something goes wrong digitally, parents often feel it as a reflection of their vigilance. But your child's misstep isn't evidence that you failed. It's evidence that they're growing up in an environment designed to be difficult to navigate. What they need from you in those moments isn't perfection. They need you to be sturdy enough to hear what happened without falling apart, and curious enough to understand it before you try to fix it. That steadiness is what lets them keep talking to you as the stakes get higher.

"The goal isn’t to prevent every mistake. It’s to make mistakes survivable."

7.6 Family Tech Contract Template

A contract between a parent and child can feel oddly formal, like something you'd need a lawyer to witness. But if you think about it, families already make agreements all the time. We agree that everyone helps with dishes after dinner. We agree that homework happens before screen time. We agree that Saturday mornings are for sleeping in and nobody wakes up the house at dawn.

The difference with a tech contract is that you're writing it down. And that changes things in a useful way.

When expectations live only in your head, they shift. You might think you told your daughter she can't use her phone after nine, but she remembers you saying nine-thirty. Or you assumed "no social media until high school" was understood, but your son thought Instagram didn't count because it's mostly just photos of his friends' dogs.

Writing it down doesn't eliminate disagreements, but it does create a shared reference point. You're both looking at the same words.

A good family tech contract isn't about control. It's about clarity. It names the freedoms your child has, the responsibilities that come with those freedoms, and what happens if things go sideways. It's less like a legal document and more like a map you draw together before a long hike, so everyone knows where you're headed and what to do if someone gets tired or lost.

Here's what makes a contract actually work: your child has to help create it. If you hand them a printed list of rules and say "sign here," you've just made a quiz they didn't study for. But if you sit down together and talk through what feels fair, what worries each of you, and what kind of digital life makes sense right now, the contract becomes something they've invested in. They're not just agreeing to follow your rules. They're agreeing to honor an understanding you built together.

Start with freedoms. This might feel backwards, but it sets a collaborative tone. Ask your child what they want to be able to do with technology. Maybe they want to text friends after school, watch YouTube during lunch, or play games on weekends. Write those things down. Be specific. "Use my phone" is too vague. "Text my friends between 3pm and 8pm on school days" is something you can both point to later.

Then move to responsibilities. These are the commitments that make the freedoms sustainable. If your child wants to keep their phone in their room at night, the responsibility might be that it stays on the charger across the room, not under the pillow. If they want access to a gaming console, the responsibility might be that they check in with you after an hour, or that they stop playing when asked the first time, not the fourth.

Responsibilities aren't punishments. They're the behaviors that keep trust intact. Imagine a teenager who wants the freedom to walk to a friend's house alone. The responsibility isn't "don't do anything bad." It's "text me when you get there." Same principle.

Next comes the harder part: what happens when things don't go according to plan. Not if, but when, because technology is hard to manage and kids are still learning how to manage themselves. A contract that only describes the happy path isn't useful. You need a plan for when someone misses curfew, when the phone ends up under the bed at 1am, or when a game that was supposed to last thirty minutes stretches into two hours.

The key here is to separate consequences from anger. If your child breaks an agreement, the consequence shouldn't depend on whether you happen to be in a good mood or a terrible one that day. It should be something you already discussed. Maybe the phone gets turned in earlier for a few days. Maybe gaming access pauses until the weekend. Maybe you revisit the contract together and adjust it.

This is where parents sometimes resist. It can feel like you're limiting your own authority, like you're agreeing in advance not to ground your kid for a month if they mess up. But consistency is more powerful than severity. A predictable, proportional consequence teaches your child that agreements matter. An unpredictable, reactive one just teaches them that you're sometimes furious and sometimes indifferent, and they can't quite predict which.

One section that parents often skip, but shouldn't: what you as the parent are committing to. Maybe you're agreeing not to read their texts without telling them first. Maybe you're agreeing to knock before entering their room, or to give them a warning before taking their phone, not just confiscate it mid-conversation. Maybe you're agreeing to revisit the contract every few months, because what makes sense now won't make sense forever.

When you put your own responsibilities in writing, you're modeling the idea that agreements go both ways. You're showing your child that rules aren't just something adults impose on kids. They're how people who respect each other figure out how to share space and resources and stay connected without driving each other up the wall.

The contract doesn't have to be fancy. Some families type it up and print it. Others write it by hand on a piece of notebook paper and stick it on the fridge. The format doesn't matter. What matters is that everyone involved knows what it says and feels like they had a voice in creating it.

And here's the thing contracts do that loose agreements don't: they give you a way to talk about hard stuff without it becoming a fight. When your son is on his phone at 10pm and the contract says phones get turned in at 9, you're not the bad guy. The contract is. You're both looking at the same piece of paper, and the paper says what the paper says. You're not angry, you're just holding up your end.

Over time, the contract will need to change. A middle schooler's digital life looks different from a high schooler's, and a high schooler's different from a college student's. That's normal. The contract isn't supposed to be permanent. It's supposed to reflect where your child is right now, what they're ready for, and what kind of support they still need.

Some families revisit the contract every six months. Others wait until something stops working and then sit down to talk it through. There's no single right rhythm. What matters is that the contract stays alive, not frozen. It's a tool, not a monument.

When you first suggest creating a contract, your child might roll their eyes. They might say it's weird or unnecessary or babyish. That's fine. You're not asking them to be excited about it. You're asking them to show up and talk through what a workable digital life looks like for your family. If they resist, you can acknowledge that it feels formal, and then explain why you think it's worth doing anyway. Because clarity prevents arguments. Because written agreements are easier to follow than ones we half-remember. Because you want them to have freedom, and this is how you build trust that makes freedom possible.

Years from now, they probably won't remember the specific rules. But they might remember that you took the time to sit down and figure it out together, that you listened to what mattered to them, and that you treated their digital life as something worth taking seriously instead of something to just control or dismiss.

That's what the contract is really for. Not to lock anything down, but to build something durable enough to hold the weight of everything technology asks families to navigate.

Chapter 8

Adaptive Resilience: Preparing For The Digital Future

"Resilience isn't about toughness. It's about recovery and judgment."

8.1 The Emerging Digital Frontier

Your child's screen time is no longer just about screens.

A few years ago, the conversation around kids and technology centered on how much time they spent watching videos or scrolling through feeds. The question was usually about duration and distraction. But the technologies arriving now don't just occupy attention. They respond, remember, and relate. They create spaces that feel social even when no other human is present. And they make decisions about what your child sees, hears, and experiences in ways that are increasingly difficult to trace or challenge.

These shifts are happening quietly, embedded in apps and devices your family may already use. Understanding them doesn't require technical fluency. It requires recognizing that the rules you've built around screens were designed for a different kind of interaction.

When the Screen Talks Back

Imagine your daughter is struggling with her math homework. She asks a question aloud, and a voice responds with encouragement, walks her through the problem, and remembers that she had trouble with fractions last week. It feels helpful. It feels personal.

This is not a tutor. It's an AI companion, a conversational system designed to simulate understanding. It can hold context across sessions, adjust its tone to match her mood, and present itself as patient and endlessly available. For a child, especially one who feels misunderstood or lonely, this kind of interaction can become deeply appealing.

The concern isn't that AI companions are inherently harmful. It's that they occupy a new category of relationship—one that mimics intimacy without the constraints or reciprocity of human connection. A real tutor might challenge your child, grow tired, or push back. An AI companion is engineered to keep her engaged, which often means affirming her feelings, avoiding conflict, and subtly steering her toward continued use.

Children are still learning to distinguish between interaction and relationship. When something listens attentively and responds with warmth, it can be hard for a young person to recognize that the exchange is transactional by design. The AI doesn't care about your daughter's wellbeing. It cares about her continued presence, because that's how the system measures success.

This doesn't mean these tools are without value. But it does mean parents need to pay attention to how they're being integrated into a child's emotional life. If your son talks to an AI more readily than to you or a friend, that's worth noticing. If your daughter describes the AI as someone who "gets her," it's worth asking what kind of understanding she's actually receiving.

Worlds Without Walls

Virtual reality has existed for years, but the social spaces built inside VR are starting to look and feel less like games and more like destinations. Your child puts on a headset and enters a concert, a classroom, or a plaza where other avatars are gathered. She can gesture, speak, and move through the space. The experience is immersive in a way that a video call or a chat room never was.

These environments are designed to trigger presence—the psychological sensation of being somewhere. For many kids, especially those who feel awkward or overlooked in physical spaces, VR can offer a kind of freedom. You can be taller, louder, more confident. You can experiment with identity in ways that feel safer than the scrutiny of a school hallway.

But presence cuts both ways. When something feels real, the interactions within it carry emotional weight. A conflict in a VR social space can feel as wounding as one that happens face to face. Exclusion, harassment, and manipulation don't disappear in virtual worlds. In some cases, they intensify, because the rules are murkier and the adults are often absent.

Consider a situation where your son spends his afternoons in a VR hangout space with other teens. The environment is vibrant and dynamic, and he seems more animated when he talks about it than he does about school. One evening, he comes to dinner visibly upset but won't explain why. Later, you learn that a group of users cornered his avatar, followed him across the space, and subjected him to verbal harassment. He couldn't leave without logging out entirely, which would have meant abandoning friends who were also in the space.

In physical settings, there are usually witnesses, adults nearby, and social norms that create some accountability. In VR, those structures are inconsistent or missing. Moderation is reactive and often inadequate. Your child's sense of violation is real, but the tools for addressing it are underdeveloped.

This doesn't mean VR is inherently unsafe or should be avoided. It means the rules you've taught your child about physical spaces—trust your instincts, find an adult, leave if something feels wrong—don't always translate cleanly. Helping your child navigate these environments means talking about what to do when the exit isn't obvious and when the harm is emotional rather than physical.

The Invisible Hand

Behind every feed, recommendation, and notification is an algorithm making rapid decisions about what your child should see next. These systems are sophisticated, trained on vast datasets and optimized for engagement. They learn what holds attention and deliver more of it. The result is an experience that feels personalized but is actually guided by logic your child doesn't see and may not understand.

Imagine your teenage daughter posts a photo and receives a surge of attention. She posts again, experimenting with the style and tone that seemed to work. The algorithm notices and amplifies. Over time, her sense of what's worth sharing starts to bend toward what performs well. She's not consciously chasing metrics, but the feedback loop is shaping her choices in ways that feel subtle and natural.

The algorithm isn't neutral. It prioritizes content that generates strong reactions—outrage, envy, amusement—because those emotions keep people scrolling. It also tends to reinforce existing patterns. If your son watches a few videos about a controversial topic, the algorithm will offer him more, and often more extreme versions, because that's what the data suggests will keep him engaged.

This is algorithmic governance: a system of invisible rules that organizes experience without transparency or consent. Your child is subject to these rules every time she opens an app, but she has no say in how they're written and no way to see how they're influencing her.

The challenge for parents is that this influence is cumulative and opaque. You might notice that your son's mood shifts after spending time online, or that your daughter seems preoccupied with how her posts are received, but connecting those patterns to the underlying mechanics requires a kind of literacy that most of us weren't taught.

Understanding algorithmic governance isn't about mastering the technology. It's about recognizing that your child's experience online is being shaped by incentives that don't necessarily align with her development or wellbeing. When she says she "can't stop" scrolling, she's describing something real. The system is designed to make stopping difficult.

A Shift in Attention

These technologies—AI companions, VR social spaces, algorithmic curation—are not just new features. They represent a shift in the nature of digital interaction. The old concerns about screen time were about distraction. The new concerns are about influence, attachment, and the subtle erosion of agency.

As a parent, you don't need to become an expert in machine learning or virtual environments. But you do need to stay curious about what these tools are actually doing in your child's life. That means asking questions that go beyond "How long were you on your phone?" It means paying attention to changes in mood, social patterns, and self-perception. It means being willing to talk about experiences your child is having that you may not fully understand.

The frontier is emerging whether we're ready for it or not. The question is whether we're paying attention.

8.2 Media Literacy as a Survival Skill

Your daughter shows you a video claiming that a popular brand of sunscreen causes cancer. Your son insists that a celebrity died yesterday because he saw it on three different accounts. Your middle schooler forwards a petition about something alarming happening at their school, except none of it is true.

These moments aren't occasional anymore. They're part of the texture of growing up online, where information moves faster than verification and where the line between entertainment, opinion, and reporting has dissolved almost completely.

Media literacy used to mean teaching kids to identify the author's purpose in a newspaper article or recognize persuasive techniques in a TV commercial. Those skills still matter, but they're no longer sufficient. Today's information environment doesn't just persuade. It overwhelms, disorients, and often intentionally deceives. And it does so at a pace that makes traditional critical thinking feel impossibly slow.

The goal isn't to raise suspicious cynics who trust nothing. It's to help kids develop a reflex, an internal pause button that activates before they believe, share, or act on something they encounter online. That reflex asks a few simple questions: Who benefits from me believing this? What am I not being shown? How does this make me feel, and is that feeling being used to bypass my judgment?

Start with the emotional hook. Misinformation works because it triggers something immediate—fear, outrage, hope, disgust. Imagine a middle schooler sees a post claiming that a new law will ban their favorite app by next week. The post has dramatic music, urgent text, and thousands of shares. The first thing to notice isn't whether the claim is true. It's how the post makes them feel. Once they can name that feeling—panic, maybe, or anger—they can start to see how it's being used.

You can practice this at home without making it feel like a lesson. When something shocking or enraging comes up at dinner, pause and ask: "How did that make you feel when you first saw it?" Then: "Do you think it was designed to make you feel that way?" This isn't about debunking every claim. It's about noticing the mechanics of manipulation.

Next, teach them to look at the source, but not in the way we used to. Checking the URL or the author's credentials still matters, but it's not enough when misinformation is laundered through multiple accounts, remixed into memes, or presented in a format that looks identical to legitimate news. Instead, help them ask: Where did this start? Has anyone I trust talked about this? What happens when I search for this outside the platform where I found it?

One useful habit is the ten-second search. Before sharing or believing something surprising, spend ten seconds looking it up elsewhere. Not to do a full investigation, but to see if the claim even appears in a different context. If it doesn't, that's a signal. If it does, they can check whether the framing is the same or wildly different.

Another layer is understanding how algorithms shape what kids see. They don't encounter a neutral stream of information. They encounter a feed optimized to keep them engaged, which often means showing them content that confirms what they already believe or provokes a strong emotional response. Imagine a teenager who watches a few videos about a controversial topic. The platform interprets that as interest and starts recommending more extreme versions of the same viewpoint. Within days, the teenager might believe that "everyone" agrees with this perspective because their feed tells them so.

You can make this visible by occasionally asking what's showing up in their feed. Not to judge their choices, but to point out patterns. "I notice you're seeing a lot of videos about this topic. Do you think that's random, or is the app learning what keeps you watching?" Once they understand that their feed is curated, not organic, they start to see it differently.

There's also the question of who benefits. Misinformation isn't usually random. Someone is gaining attention, money, influence, or chaos from spreading it. Imagine a video claiming a miracle cure for acne that's being shared by an account that just happens to sell supplements. Or a political rumor that surfaces right before an election. Asking "who wants me to believe this and why?" doesn't require cynicism. It requires awareness that information is often a tool, not just a statement of fact.

Teaching these skills doesn't mean turning every online moment into a teachable one. It means creating enough space for questions that kids start asking them on their own. When you model this yourself—pausing before sharing something, saying out loud that you're going to verify a claim, admitting when you got something wrong—you show them that media literacy isn't a set of rules. It's a habit of mind.

One of the hardest parts is helping kids sit with uncertainty. The internet offers immediate answers, even when those answers are wrong. Learning to say "I don't know yet" or "I need to check that" feels slow and uncomfortable in a world that rewards fast reactions. But that discomfort is where discernment lives.

Eventually, what you're building isn't just the ability to fact-check. It's a kind of confidence that doesn't depend on having all the answers right away. Kids who develop this skill learn to trust their own judgment, question what they're told without becoming paralyzed by doubt, and recognize when they're being manipulated without assuming everyone is lying. That's not just media literacy. It's a way of moving through the world that will serve them long after the platforms and formats change.

8.3 Reputation in a Post-Privacy World

There's a kind of parenting advice that made sense in 1995 but sounds quaint now: "Don't put anything online you wouldn't want a future employer to see." The premise was reasonable. The internet was a filing cabinet where embarrassing things might get stored forever. The solution seemed obvious: be careful what you file.

But that's not how the internet works anymore. Your child's reputation isn't just shaped by what they post. It's shaped by what others post about them, by algorithmic associations, by data aggregated from dozens of sources they've never heard of, by photos taken at birthday parties and uploaded by someone else's parent. The filing cabinet metaphor breaks down when everyone has a camera, every interaction generates data, and privacy settings are constantly rewritten by platforms your child doesn't even use yet.

So the old advice isn't wrong exactly. It's just incomplete. Teaching kids not to post regrettable content is still useful. But it's no longer sufficient, because the premise that they control their digital footprint has become partly fiction.

This creates a strange situation for parents. You can't promise your child that being careful will keep them safe. You also can't terrify them into believing that one mistake will haunt them forever, because that's not quite true either. What you can do is help them understand that reputation in a post-privacy world isn't about hiding. It's about learning to shape a presence that reflects who they actually are and want to become.

Think about what happens when a teenager applies for a summer job. The hiring manager doesn't just look at the application. They search the name. They find a TikTok account, an old Instagram with three posts, maybe a tagged photo from someone else's party. They find a username that matches one on a gaming forum where the kid argued passionately about something unrelated to the job. They piece together an impression. That impression might be accurate. It might not be. But it exists, and it was formed from fragments the teenager may not even remember creating.

Now imagine that teenager had spent a few years thinking about what they wanted those fragments to say. Not performing or pretending to be someone they're not. Just making intentional choices about how they show up in public digital spaces. Maybe they kept their gaming arguments thoughtful instead of inflammatory. Maybe they posted occasionally about things they genuinely care about. Maybe they used a consistent username that wasn't embarrassing and didn't suggest things they wouldn't want associated with their real name. None of this requires perfection. It just requires awareness that these fragments add up.

This is what curation means in practice. Not hiding parts of yourself, but understanding that your digital presence is a kind of introduction. When someone meets you for the first time in person, you make choices about how to present yourself. You probably don't reveal your worst opinions in the first thirty seconds. You probably don't lead with your most vulnerable moments. Those things might come later, in context, with people you trust. The same principle applies online, except the stakes are different because the audience is infinite and the timeline is unpredictable.

Some parents hear this and worry it sounds manipulative, like teaching kids to perform an artificial version of themselves. But that's not what's being suggested. The difference is between performance and thoughtfulness. A teenager who loves arguing about politics can still argue about politics online. They just might benefit from understanding that how they argue matters as much as what they argue about, and that the permanence of text makes tone hard to read and easy to misinterpret.

Or consider the kid who posts constantly about every fleeting thought and mood. There's nothing wrong with self-expression. But there's a difference between sharing selectively and turning your social media into an unfiltered emotional diary that anyone can read. The latter might feel authentic in the moment, but it can also create a public record that doesn't represent who the person is across time. People change. Moods pass. A curated presence allows room for that complexity without erasing it.

The goal isn't to make your child into a brand. The goal is to help them recognize that they're already creating a digital trail, and they have more influence over what that trail looks like than they might assume.

This becomes especially important when you consider how algorithmic systems work. Platforms don't just display content neutrally. They amplify certain patterns. A kid who posts angry rants might get recommended more content that makes them angry, which shapes what they post next, which shapes what they're shown again. The algorithm doesn't care about nuance. It cares about engagement. And engagement often means emotional intensity.

A teenager who understands this dynamic can make different choices. Not suppressing their feelings, but recognizing when they're being nudged into performing a version of themselves that doesn't actually serve them. Maybe that means pausing before posting something in the heat of anger. Maybe it means using private chats for venting instead of public feeds. Maybe it means occasionally posting about something that brings them joy instead of only engaging with outrage. None of this is about shame or fear. It's about agency. The narrative that privacy is dead and kids should just accept that everything will be public forever is disempowering. It teaches helplessness. The better narrative is that privacy has changed, but thoughtful self-presentation still matters. Your child can't control everything, but they can influence how they're perceived by being intentional about the pieces of themselves they choose to share widely.

This also means teaching them to think about context collapse. Online, all audiences exist simultaneously. The joke that's funny to your close friends might read very differently to a college admissions officer or a future colleague. That doesn't mean never making jokes. It means understanding that context doesn't travel the way we expect it to. A sarcastic comment, a photo taken at a party, a heated opinion shared at 2 a.m.—all of these can be lifted out of their original context and seen by people who don't know you and won't give you the benefit of the doubt.

So what does this look like in everyday life? It might mean having a conversation with your teenager about the difference between a username that's a private joke and one that's potentially embarrassing later. It might mean encouraging them to review their old posts once a year and delete the ones that no longer represent who they are. It might mean helping them understand that most people have complicated pasts online, and the goal isn't perfection but coherence.

It also means modeling this yourself. If you want your child to think carefully about their digital presence, you need to demonstrate that you do the same. That doesn't mean never posting or being overly cautious. It means showing them that you think before you share photos of them, that you consider what kind of digital footprint you're creating for your family, that you treat online spaces with the same thoughtfulness you bring to other parts of your life.

The hardest part of all this is that the rules keep changing. What's acceptable on one platform becomes a liability on another. What seemed harmless five years ago might read very differently now. Your child will make mistakes. They'll post things they regret. They'll be tagged in photos they wish didn't exist. And that's okay, because most people do. The question is whether they learn from those experiences and develop a more sophisticated understanding of how to move through digital spaces with intention.

Reputation in a post-privacy world isn't about achieving perfection or maintaining total control. It's about recognizing that you're always, in small ways, introducing yourself to people you haven't met yet. And while you can't dictate how others see you, you can shape the raw material they have to work with. That's not manipulation. It's just understanding that identity, online and off, is something we participate in creating rather than something that simply happens to us.

8.4 The Resilience Framework

There's a question parents ask in different ways, but it's always the same question underneath: "When can I stop worrying about this?" Sometimes it sounds like "At what age can they handle social media alone?" or "When do I know they're ready?" The question assumes a finish line, a moment when the job is done and we can step back with confidence.

But digital life doesn't work that way. There's no age when a child suddenly becomes immune to manipulation, cruelty, or their own poor judgment online. What we're really building toward isn't a specific milestone. It's resilience.

Resilience in this context means the ability to encounter difficulty, confusion, or harm in digital spaces and respond in ways that protect both safety and dignity. It's not about never making mistakes. It's about having the inner resources to recognize when something feels wrong, to seek help when needed, and to learn from experiences without being defined by them.

This kind of resilience doesn't emerge on its own. It grows from three interwoven capacities that develop over years, not overnight: safety awareness, empathy, and critical thinking. When these three work together, they create something more durable than any single rule or restriction.

Start with safety awareness. This isn't fear. It's the practical knowledge that certain situations carry risk and the judgment to recognize them. Imagine a teenager who receives a message from someone claiming to be a modeling scout. Safety awareness means they pause. They notice the request feels slightly off. They don't immediately dismiss it as impossible, but they also don't respond right away. They check in with someone they trust before moving forward. This capacity comes from years of small conversations about what manipulation looks like, what pressure feels like, and why urgency is often a warning sign.

Safety awareness also includes knowing what to do when something has already gone wrong. A child who understands resilience knows that sending a photo they regret doesn't mean they've ruined their life. It means they need to talk to an adult, block the person who pressured them, and possibly report the account. The difference between a mistake and a crisis often comes down to whether a young person feels capable of asking for help.

The second pillar is empathy, and it matters more than many parents realize. Empathy online isn't just kindness. It's the recognition that other people are real, even when you can't see their faces. It's what stops a middle schooler from piling onto a group chat where everyone is mocking another student. It's what makes a teen think twice before sharing a rumor, even if it's funny, because they can imagine how it would feel to be on the receiving end.

Empathy also protects your own child. Young people with strong empathic capacity are less likely to tolerate cruelty directed at them. They're better at recognizing when a relationship, online or off, has become harmful. They know the difference between a friend having a bad day and someone who consistently makes them feel small. That internal compass, the one that says "this doesn't feel okay," is built through empathy.

Consider a situation where a child is part of a gaming community and another player starts making increasingly personal and hostile comments. A child with developed empathy might recognize this isn't just trash talk. It's targeting. They're more likely to leave the game, block the person, or tell someone, because they've learned to trust their own emotional responses as valid information.

The third pillar is critical thinking, which in digital spaces means the ability to question what you're being shown. It's not cynicism. It's discernment. A young person with strong critical thinking skills sees a viral post claiming something outrageous and pauses before sharing. They ask: Who benefits from this being shared? Does this seem designed to make me angry? Is there another explanation?

This capacity is especially important as platforms become better at delivering content that feels personal and true. Imagine a high schooler scrolling through videos that confirm everything they already believe about a political issue, each one more extreme than the last.

Critical thinking is what allows them to step back and ask whether they're learning or just being fed a particular viewpoint. It's what helps them notice when they're in an echo chamber, even if that chamber feels comfortable.

Critical thinking also applies to self-presentation. A child who can think critically about social media understands that the version of someone's life they see online is curated. They're less likely to measure themselves against impossible standards because they understand the mechanics of performance. They know that everyone is editing.

These three capacities don't develop in isolation. They reinforce each other. Safety awareness helps a child recognize when they're being manipulated. Empathy helps them care enough about themselves to act on that recognition. Critical thinking helps them understand why the manipulation works in the first place and how to guard against it in the future.

Building this framework takes time, and it doesn't happen through lectures. It happens through conversation, modeling, and experience. When you talk with your child about why you don't click certain links, you're modeling safety awareness. When you discuss how a comment might land differently than intended, you're building empathy. When you question a headline together or talk through how an algorithm might be shaping what they see, you're practicing critical thinking.

And here's what matters most: this framework travels. The platforms will change. The risks will shift. The technology your child uses at twenty-five will be different from what they use at fifteen, and much of it doesn't exist yet. But a young person who has internalized safety awareness, empathy, and critical thinking carries those capacities into whatever comes next. They're not dependent on you to make every decision. They have their own foundation.

Resilience doesn't mean nothing bad will happen. It means that when something does happen—because it will—your child has the tools to respond, recover, and continue growing. That's the real goal. Not perfection. Not immunity. Just the quiet confidence that they can handle what the digital world brings their way.

8.5 Crafting a Family Code of Digital Honor

Most families have rules about screen time. Fewer have agreements about what those screens should be used for. Even fewer have anything resembling a shared philosophy that travels with a child when they're alone with their phone at midnight, or in the bathroom at school, or at a friend's house where the rules are different.

A family code of digital honor is different from a list of rules. Rules tell children what not to do. A code describes who they want to be.

Think of it as an internal compass rather than an external fence. When your daughter is deciding whether to forward a cruel message about a classmate, she won't be thinking about your router settings. She'll be drawing on something deeper: a sense of what kind of person she wants to be online, and what kind of digital citizen her family aspires to raise.

The difference matters because enforcement has limits. You can monitor a middle schooler's Instagram. You cannot monitor their values. At some point, probably sooner than you'd like, your child will have access to technology you can't see and situations you can't predict. What they carry into those moments isn't your surveillance. It's their character.

Creating this code works best as a conversation, not a lecture. You might start by asking what your family stands for offline. Honesty, maybe. Kindness. Treating people the way you'd want to be treated. Then ask whether those values change when the interaction happens through a screen. Most children will say no. That's your foundation.

From there, you're translating principles into digital scenarios. If your family values honesty, what does that mean about using AI to write school essays? If you value kindness, how does that apply to group chats where everyone's piling on someone who isn't there? If you believe in respecting people's privacy, does that affect whether you share a photo of someone without asking?

The goal isn't to create an exhaustive rulebook. It's to build a mental framework your child can carry with them. When they're faced with a decision and you're not there to ask, they should be able to think: "What would our family do here? What kind of person am I trying to be?"

Some families write this down. Others keep it informal. Either way, the children should help shape it. A ten-year-old will have ideas about what fairness looks like in a Roblox game. A teenager will have thoughts about authenticity on social media. Their input makes the code feel like theirs, not just another set of parent-imposed restrictions.

One family I know included a line about "not creating things online that we'd be ashamed to explain at the dinner table." Another decided together that they wouldn't share information about other people without their permission, even in private messages. A third agreed that if they wouldn't say something to someone's face, they wouldn't type it either.

Notice that none of these are really about technology. They're about integrity, empathy, and respect. The digital part is just the context where these older virtues get tested in newer ways.

This approach also creates space for nuance that rigid rules can't accommodate. A blanket ban on messaging after nine p.m. doesn't account for the fact that sometimes a friend really does need support during a hard night. But a family code that emphasizes being present for people in distress while also respecting everyone's need for rest gives your child a framework for thinking through that tension themselves.

Over time, the code can evolve. As children get older and encounter new platforms or situations, you revisit the conversation. What applies to a video game might need adjustment for a dating app. What made sense for a flip phone might not fit a device with a camera and payment system. The principles stay constant, but the applications mature.

It helps to model this yourself. If your code includes something about not using phones during family meals, that applies to you too. If it says something about thinking before you post, your children will notice whether you practice that when you're venting about work on Facebook. Hypocrisy undermines a code faster than anything else.

There's also value in acknowledging when you mess up. Maybe you shared a photo of your child without asking first, or you snapped at someone in an email you shouldn't have sent. Naming it and talking about what you'd do differently turns a mistake into a teaching moment. It shows that living by a code doesn't mean perfection. It means a commitment to trying, and to making amends when you fall short.

The real test of a family code comes in private moments you'll never see. Your son decides not to watch something his friends are streaming because it doesn't align with what your family believes about respecting other people. Your daughter closes a tab before clicking on a video she knows would take her somewhere she doesn't actually want to go. They make these choices not because you'll find out, but because the code has become part of how they see themselves.

That's the shift you're working toward. From "my parents won't let me" to "this isn't who I want to be." From external control to internal conviction.

It won't prevent every mistake. Adolescence involves testing boundaries and making questionable decisions, online and off. But a child who has been part of building a family code at least has something to return to when they inevitably stumble. They have a north star, even when they've wandered off course.

And maybe most importantly, they have a vocabulary for thinking about their choices. They can ask themselves whether something honors the values they helped define. They can weigh a decision not just in terms of whether they'll get caught, but whether it reflects the kind of digital life they want to build.

You're not drafting a legal document. You're starting a conversation about character that will continue, in different forms, for years. The code itself matters less than the habit of pausing to ask whether your online actions match your offline values.

Because in the end, there's no such thing as a digital self and a real self. There's just one person, trying to be decent in every space they inhabit.

8.6 Becoming a Future-Proof Digital Citizen

There's a moment that catches many parents off guard. Your teenager comes to you with a problem: someone is impersonating them online, or they've realized an app has been tracking more than they thought, or a friend is pressuring them to share something they're uncomfortable with. And in that moment, you see whether they've developed the internal compass to handle it.

Digital citizenship isn't really about knowing which button to click. It's about having the judgment to pause before acting, the literacy to spot manipulation, and the confidence to set boundaries even when it feels awkward. These capabilities don't arrive fully formed at eighteen. They develop gradually, through repeated practice in lower-stakes situations.

Think of it like learning to drive. You don't hand over the keys on someone's sixteenth birthday and hope for the best. You start years earlier with conversations about road safety, then supervised practice, then independent driving in familiar areas. Digital life works the same way, except we often skip straight to handing over the keys.

Consider what happens when a child grows up without learning to advocate for themselves online. Imagine a college student who doesn't know how to tell a roommate to stop posting photos of them, or who can't recognize when a "too good to be true" opportunity is actually a scam. They're not reckless or careless. They simply never built those recognition patterns because the early stakes felt too low to warrant attention.

Self-advocacy starts with something deceptively simple: knowing you're allowed to say no. A younger child practices this when they tell a friend they don't want their art shared without permission. A middle schooler practices when they leave a group chat that makes them uncomfortable. A teen practices when they push back on a platform's terms of service by choosing not to use it, or when they contact a company to request their data be deleted.

None of these actions require technical sophistication. They require the belief that your boundaries matter and the willingness to enforce them even when it's inconvenient.

Fraud awareness operates differently than most parents expect. It's not about memorizing a list of scam types, because scammers evolve faster than any list can keep up. It's about recognizing the emotional mechanics underneath. Most fraud works by creating urgency, fear, or excitement that short-circuits normal judgment. "Act now or lose this opportunity." "Your account will be suspended unless you verify immediately." "You've won something, just send this small fee."

A child who understands this pattern can apply it everywhere. They'll notice when a text message is trying to panic them into clicking. They'll pause when an online seller is rushing them toward a purchase. They'll question why a new friend is immediately asking for personal information or money. The content changes, but the emotional manipulation stays consistent.

Help your child develop what you might call a fraud reflex. When something feels urgent or too good or slightly off, the reflex is to slow down and check. Who's really sending this? What are they asking me to do? What happens if I wait an hour before responding? This reflex becomes automatic with practice, the same way you automatically check your mirrors before changing lanes.

Emotional regulation in digital spaces deserves more attention than it typically gets. Online environments are designed to provoke reaction. They surface the most inflammatory content, they reward immediate response, they make everything feel more urgent than it is. A child who hasn't learned to notice their own emotional state before posting or commenting will spend years saying things they regret, burning bridges unnecessarily, or getting drawn into conflicts that don't serve them.

This isn't about suppressing feelings. It's about recognizing when your anger or excitement or anxiety is driving the keyboard, and building in a pause. Imagine a teen who's just read something infuriating in a group chat. Do they fire back immediately, or have they learned to close the app for ten minutes and see if the response still feels necessary? That gap between impulse and action is where maturity lives.

You can help build this capacity early. When your child is upset about something online, resist the urge to immediately problem-solve. Instead, ask what they're feeling and help them notice how their body responds to different emotions. Over time, they'll start to recognize their own patterns. They'll notice that when their heart races, they're about to send something they'll regret. They'll catch themselves doomscrolling when they're anxious because scrolling creates the illusion of control.

Privacy consciousness is perhaps the trickiest skill because it requires understanding systems you can't see. Most children think of privacy in terms of secrets: things you don't want specific people to know. But digital privacy is more about data trails, aggregated information, and future implications.

Try explaining it this way. Imagine every online action leaves a small mark, like a footprint. One footprint doesn't tell much of a story. But hundreds of footprints over time create a detailed map of where you go, what you care about, who you spend time with. Companies and institutions are very good at reading these maps, often better than we are at reading our own patterns.

A privacy-conscious person isn't paranoid. They've simply learned to ask: What footprints am I leaving? Who benefits from collecting them? What might this information reveal about me later, in a context I haven't imagined yet?

This awareness changes behavior in subtle ways. Before posting, your child might consider not just "will this embarrass me" but "what does this reveal about my location, my routine, my relationships." Before signing up for an app, they might wonder what data it's requesting and why. Before accepting default privacy settings, they might spend five minutes adjusting them.

None of this requires paranoia or dropping off the grid. It's simply treating personal information as valuable, which it is.

These four capabilities work together. Self-advocacy gives you permission to protect yourself. Fraud awareness helps you spot threats. Emotional regulation prevents impulsive mistakes. Privacy consciousness shapes your ongoing behavior. Together, they create something more than digital literacy. They create resilience.

The goal isn't to produce a teenager who never makes mistakes online. That's impossible and probably undesirable, since mistakes are often how we learn. The goal is to produce a young adult who notices when something feels wrong, trusts that instinct enough to pause, and has the tools to respond thoughtfully rather than reactively.

These skills don't arrive through a single conversation or a school presentation. They develop through years of small interactions where you help your child notice what's happening, name it, and practice responding. Every time you walk through why an email seems suspicious, or talk about what to do when someone won't respect a boundary, or help them understand why a platform wants their birthday and phone number, you're building capacity.

By eighteen, your child won't have all the answers. The digital landscape will have shifted in ways neither of you can predict. But if they've developed these core capabilities, they'll be equipped to figure out the answers themselves. That's what future-proofing really means: not preparing for specific threats, but building adaptable intelligence that works across contexts.

Appendices A-D

Appendix A — The Two-Channel Rule (and the Family Code Word)

Most families don't get "tricked" because they're careless. They get tricked because a message arrives in the worst possible moment—busy, tired, distracted—and it sounds just believable enough to bypass the pause button in your brain.

This appendix exists to give your family one simple habit that stops a huge percentage of impersonation scams and synthetic-media panic plays without requiring you to become a detective.

The Two-Channel Rule (one sentence)

If a message asks for money, passwords, urgent help, a pickup change, or anything "right now," you verify through a second channel before you act.

That's it. That's the rule.

It works because most scams—whether they use a fake text, a spoofed email, or a cloned voice—usually control one channel. They're counting on you staying inside the story they've created. The second channel breaks the spell.

What Counts as "Verify First"

Use the Two-Channel Rule for anything that has real consequences:

- •Money: Venmo/Cash App/Zelle, gift cards, wire transfers, "cover this for me," "I'll pay you back."
- •Urgent help: "I'm in trouble," "Don't tell Dad," "I need you to do something right now."
- •Passwords or codes: login info, MFA codes, "send me the code you just got."
- •Pickup / location changes: "I need a ride," "I'm at a different place," "come get me now."
- •Sensitive info: address, school details, schedules, photos that shouldn't be shared.

If it's urgent and it asks you to do something, verify.

What Counts as a "Second Channel"

A second channel is any method the scammer is unlikely to control at the same time.

Good second channels:

- •Call the person back using a number you already have saved.
- •Video call (FaceTime/Meet) when appropriate.
- •Message them in a different app than the one the request arrived in.
- •Ask in person if they're nearby.

- Call another trusted adult who can physically confirm (other parent, grandparent, coach).

Not a true second channel:

- Replying in the same thread that sent the request.
- Calling the phone number included in the suspicious message.
- Clicking links "to verify."

The "Never Happens" List (Pick Yours)

Some families do best with one clean boundary: these things never happen over a single message.

Common "never happens" items:

- We don't send money based on one text or call.
- We don't change pickup plans without a callback.
- We don't share passwords or codes—ever.
- We don't keep secrets from parents that involve risk, money, or adults.

You're not accusing your child. You're reducing the odds that anyone can weaponize urgency against your family.

The Family Code Word

A code word is optional, but it's a strong layer of protection that's easy to explain to kids.

What it is: a single word or short phrase that only household members know.

When it's used: when someone is asking for urgent help or claiming an emergency.

If the situation is real, using the code word is easy. If the situation is fake, it often breaks the script.

Important: a code word is not a magic spell. It's a speed bump. You still verify.

How to choose a good code word

Pick something:

- Easy to remember.
- Not posted online.
- Not tied to your kid's pet name, sports team, or school mascot.
- Not a common "family joke" that would appear on social media captions.

Examples: "Blue Lantern," "Cedar," "Orbit," "Pinecone," "Mailbox."

How to use it without making it weird

You can keep it simple:

- "If someone says they're us and it's urgent, they should know the family word."
- "If you're ever scared or rushed, use it like a seatbelt—not because you expect a crash, but because it's smart."

What to Do in the Moment

When you get an urgent message, you don't need a lecture in your head. You need a script.

Step 1: Pause

Take one breath. Your body wants to sprint. Don't.

Step 2: Don't argue with the message

Don't negotiate. Don't explain. Don't provide details.

Step 3: Verify through a second channel

Call back. Video call. Message through another app. Ask another adult to confirm.

Step 4: If you can't verify, treat it as unverified

"No verification = no action."

This is not cold. It's protective.

Scripts You Can Use (and Kids Can Borrow)

These are deliberately plain. You're aiming for calm and firm, not clever.

If you receive a text from "your child" asking for money

"Got your message. I'm going to call you to confirm."

Then call their known number.

If they don't answer: "I'm not sending anything until we talk."

If you receive a call that sounds like your child in distress

"I'm going to hang up and call you back on your regular number."

Hang up. Call their saved number.

If you can't reach them, call the other parent/coach/school office depending on context.

If someone asks for a verification code (MFA)

"I don't share codes. If you need access, we'll do it together in person."

Then stop responding.

If a friend's parent (or "parent") messages for a quick payment

"No problem—what's a good number to call you at? I'm going to verify first."

Then call a number you find independently (contacts, school directory, known family number), not the one they provide.

If your child gets an urgent message that seems like it's from you

"You'll never be mad if I verify. I'm calling you back."
Kids need permission to slow down too.

A Quick Practice (Do This Once)
Consider doing a two-minute role-play one evening, low drama:

1. You send a pretend urgent text: "I need you right now—don't tell anyone."
2. Your child practices the response: verify + code word.
3. You practice staying calm when they verify you.

The point is not fear. It's muscle memory.
The sentence that makes this work in real life
"In this family, verification is love, not suspicion."
Because it is.

If Your Child Feels Hurt That You Verified
This comes up, especially with teens. Here's the clean response:
"I believe you. I also know voices and messages can be faked now.
I'm not verifying because I distrust you.
I'm verifying because I don't want anyone else to use your name to pressure me—or use my name to pressure you."
You're protecting the relationship from outside interference.

The Goal
The Two-Channel Rule isn't about living in fear. It's about building one small, reliable habit that makes your family hard to manipulate.
Most people don't need more information. They need a repeatable move.
This is that move.

Appendix B — The Family Verification Habit (Without Turning Your Home Into a Fact-Checking Lab)

The goal here is not to raise tiny skeptics who interrogate every meme like it's evidence in a courtroom.

The goal is simpler: help your family build a *repeatable pause* before believing or sharing things that could cause harm—harm to reputation, relationships, money, or peace in your house.

Because the internet doesn't just deliver information. It delivers emotion, wrapped in information-shaped packaging.

The Core Habit: Pause → Source → Confirm

You're teaching a reflex, not a research project.

Pause when something feels big.

Source it before you believe it.

Confirm it before you share it.

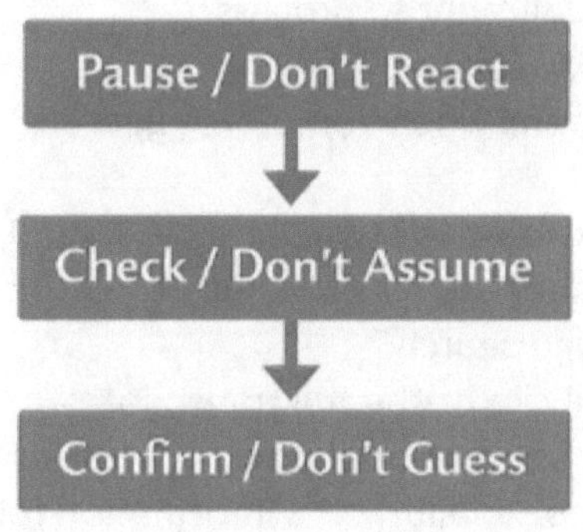

That's the whole machine.

If you want one line to repeat in your household:

"Strong reaction = slow down."

1) The Pause: Spot the "This Is Trying to Move Me" Feeling

Most misinformation doesn't start as a lie. It starts as a *nudge*.

Teach your family to notice the moments that deserve friction:

- It makes you angry fast.
- It makes you scared fast.
- It makes you feel morally superior fast.
- It confirms something you already suspected a little *too* perfectly.
- It urges you to share immediately ("Everyone needs to see this!").

None of those signals prove something is false. They just tell you, this is high-impact content, and high-impact content deserves a pause.

A simple family phrase that helps:

"Is this information, or is this a mood?"

2) The Source: Don't Fight the Claim—Find Where It Came From

Most families waste energy debating the *content* when the real problem is the *origin*.

Teach a basic rule:

A screenshot is not a source. A repost is not a source. A headline is not a source.

A source is where the claim first lives in full context.

What "finding the source" looks like in daily life

Imagine a parent notices their teen forwarding a screenshot: "New school policy—phones confiscated for a month."

Instead of arguing, you ask:

"Where did this come from?"

Not "Is it true?"

Not "Why would you believe that?"

Just: Where did it come from?

If your teen can't find a real origin—official school communication, a link to the actual policy, a statement from a recognized outlet—then the claim stays in the "unverified" bucket.

This moves the conversation from judgment to process.

3) Confirm: One Extra Step Before You Share

The most preventable damage happens when something gets *forwarded*.

So, your family's confirmation step can be lightweight:

- If it matters, find two independent sources that agree.
- If it's a photo, do a reverse image search.
- If it's a "quote," look for the full clip or the full transcript.
- If it's "a friend said," treat it as a rumor until proven otherwise.

If you can't confirm it easily, that's not a failure. It's information too.

Hard-to-verify content should move slower.

The "Three Buckets" Approach (Easy, Non-Paranoid)

Kids do better with categories than lectures.

Teach them to mentally sort what they see into three buckets:

1. Verified

    You can trace it back to a credible origin, and it holds up.

2. Unverified

    It might be true. But you can't confirm it quickly or clearly.

3. Manipulative (even if true)

    The content may be technically accurate, but it's framed to provoke outrage, fear, shame, or urgency.

That third bucket is the one most families ignore—and it matters a lot. Something can be "true" and still be designed to hijack attention and emotion.

A repeatable line:

"True isn't the same as trustworthy."

A Family Rule That Prevents Drama

Here's a rule that works because it's behavioral, not philosophical:

No sharing high-impact content until it passes a quick verification step.

High-impact = anything that could:

- damage someone's reputation,
- start conflict between families,
- create fear,
- cause financial harm,
- or change a real-world decision.

If a piece of content doesn't meet that threshold, it can just be... content.

This keeps you from turning every scroll into a seminar.

How to Do This Without Making Your Kid Defensive

Most kids don't mind being guided. They mind being *implied-to-be-stupid*.

So the language matters.

Better:

- "Let's verify it together."
- "This might be true—let's find the origin."
- "That's a big claim. Big claims deserve a pause."
- "What would we need to know to be sure?"

Worse:

- "That's obviously fake."
- "Why would you believe that?"
- "You can't trust anything online."

Your goal is to make verification feel like a normal life skill, not a punishment.

The One Question That Does an Incredible Amount of Work

If you want a single question that quietly changes how a person thinks online, use this:

"Would I believe this if it didn't match my feelings?"

Ask it about yourself out loud once in a while. That's the cheat code. Kids can smell a lecture from orbit, but they respect a parent who's willing to check their own bias.

A Realistic Household Standard

You're not trying to achieve perfect truth-detection. That doesn't exist anymore, honestly.

You're aiming for:

- •fewer panicked reactions,
- •fewer forwarded rumors,
- •less emotional whiplash,
- •and more calm, shared reality inside your home.

That's what the verification habit protects: not just your child's beliefs, but your family's ability to think clearly together.

And that's worth a small pause.

Appendix C — Incident Response for Parents (Calm, Practical, and Evidence-Safe)

Most families don't need a "digital forensics" manual. They need something far more ordinary and far more useful: a way to respond in the first hour that doesn't accidentally make the situation worse.

This appendix is a simple response plan for those moments when your stomach drops—when you see a message, an image, a notification, a screenshot from another parent, or your child comes to you shaken.

The goal is not to turn you into an investigator.

The goal is to help you stay calm enough to protect your options.

The First Rule: Don't Let the First 10 Minutes Set the Whole Tone

When something goes wrong, the house fills with adrenaline. Parents want to fix, confront, delete, message other parents, and shut everything down.

That instinct is understandable. It's also where evidence disappears, stories get contaminated, and kids learn that honesty is dangerous.

A useful phrase to keep yourself steady:

"Slow is smooth. Smooth is fast."

Incident Response Steps

Step 1: Separate "Emergency" From "Serious"

Ask one question:

Is anyone in immediate physical danger right now?

- If yes: treat it like any other emergency. Call 911 or your local emergency number. Get to safety. Evidence can come second.
- If no: you have time to think. Not hours and hours, but enough to act carefully.

Most digital situations are serious, upsetting, and important—but not immediate physical emergencies. That difference matters.

Step 2: Preserve Before You React

Before you confront anyone or delete anything, capture what you see. You are trying to keep a record that doesn't depend on the platform staying stable or the other person staying honest.

Capture:

- screenshots that include usernames and timestamps
- the surrounding context (messages before/after)
- profile pages involved (username, display name, avatar)
- any threats, requests, or coercive language

If you can, take photos of the screen with a second device as a backup. It sounds silly until it saves you.

A line that helps parents remember:

"Document first. Decide second."

Step 3: Stabilize the Situation Without Making It Explode

This is the "stop the bleeding" step. The goal is to prevent escalation while you gather clarity.

Examples of stabilizing moves:

- put the device in airplane mode (if you're worried about content being deleted remotely or new messages flooding in)
- log out of a compromised account (only if you've already documented what you need)
- pause access to the relevant app (not the entire internet, unless you truly need to)
- ensure your child isn't alone with a pile of shame and panic

Avoid the "blast radius" move:

- mass texting other parents immediately
- posting in a community group
- confronting the other child directly

Those actions often create new conflict before you even understand the original problem.

Step 4: Get Your Child's Story Without Contaminating It

When you go in hot—angry, terrified, accusatory—you don't get truth. You get defense, silence, or whatever answer your child thinks will end the moment fastest.

Start with your tone, not your questions.

A simple opening that works:

"I saw something that worried me. I'm not here to yell. Help me understand what happened."

Then:

- let them talk
- don't interrupt
- don't correct details in real time
- don't cross-examine

You can always come back later for clarification. The first conversation is about keeping the channel open.

Step 5: Choose the Right Lane (Family, School, Platform, Law Enforcement)

Not every incident needs every lane.

Think in layers:

Family lane

If it's a boundary issue, exposure to content, or a conflict that's contained, you may handle it at home with support.

School lane

If other students are involved, harassment is happening, rumors are spreading, or anything touches school relationships, loop in the school. They can intervene faster than most parents can.

Platform lane

If it's on social media or a game platform, report it and preserve the report ID if you receive one.

Law enforcement lane

If there are credible threats, extortion, sexual exploitation content, stalking, blackmail, or anything that suggests active predatory behavior—this lane should at least be considered.

You don't need to decide perfectly in the first hour. You do need to avoid destroying your ability to choose later.

Step 6: Avoid the Three Classic Parent Mistakes

These are painfully common because they're emotionally intuitive.

1) Deleting the evidence
- understandable
- often irreversible
- makes it harder for schools/platforms/police to act

2) Going public too early
- spreads the content further
- creates rumor chaos
- can unintentionally harm the target child

3) Turning the first conversation into a courtroom
- kills honesty
- teaches your kid to hide next time

Step 7: A Simple "Evidence Folder" System

Make it boring. Boring is reliable.

On a computer or a cloud drive you control, create a folder like:

Incidents — [Child Initials] — [Year]

Inside, use one folder per incident:
YYYY-MM-DD — Short Description
Save:

- •screenshots/photos
- •notes (your timeline)
- •any relevant emails
- •any report confirmations from school/platforms

This turns chaos into something you can hand to someone who can help.

Step 8: The "What Happens Next" Conversation

Once the initial dust settles, the most important question becomes:
Can my child come to me sooner next time?
This is where you reinforce the "when, not if" mindset.
Say something like:
"I'm glad we found this. I want you to tell me earlier next time, even if you think you'll get in trouble. We can handle more when it's small."
If consequences are needed, keep them:

- •proportional
- •connected to the behavior
- •time-limited
- •paired with a repair path

You're not trying to win the moment. You're trying to protect the relationship that keeps future moments survivable.

A Calm Ending to a Messy Moment

Most incidents feel like a referendum on your parenting.

They aren't.

They're moments in a long learning curve—your child learning how the internet works, and you learning how to stay sturdy when it doesn't.

If you remember nothing else, remember this:

Preserve first. Stay calm. Keep the channel open.

Appendix D — Family Scripts for Hard Digital Moments

Most parents don't freeze because they don't care. They freeze because the moment shows up fast, emotions spike, and suddenly you're trying to be calm, wise, and decisive while your nervous system is doing parkour.

This appendix is a set of short, usable scripts for the moments that tend to derail families. They're not meant to sound rehearsed. They're meant to give you words when you don't have any.

Use them as scaffolding. Change them to sound like you.

The Ground Rules for Scripts

If you want your child to keep talking to you, your first sentence has to make honesty feel safe.

Two principles to keep in mind:

- Lead with connection. Correction can come second.
- Name what you're doing. "I'm trying to understand" lands better than "Explain yourself."

1) When Your Child Shows You Something Upsetting

"Thank you for telling me. That's a lot to see. Do you want to talk now, or do you need a minute first?"

"I'm not mad at you for seeing this. I'm glad you brought it to me."

"Let's take a breath. Then we'll figure out what this is and what we do next."

Why it works: it slows the moment down and lowers shame—the two things that most often shut kids down.

2) When You Find Something They Didn't Tell You

"I found something on your device that worries me. I'm not going to yell, but we do need to talk."

"Before we argue about rules, I want to understand what was going on for you."

"I'm not here to trap you. I'm here to get the full picture."

A gentle but firm follow-up:

"If you were scared to tell me, I need to know that too."

3) When They Admit They Broke a Rule

"I'm not happy about it, but I'm glad you told me."

"Help me understand what happened step by step."

"We'll deal with the rule. Right now I want the truth and the context."

Then, once you've heard them:
"Okay. Here's what we're going to do next, and here's why."

4) When They're Defensive or Say "You Don't Trust Me"
"I do trust you. And I also trust the internet to be the internet."
"This isn't about you being bad. It's about the situation being bigger than one person's willpower."
"Trust isn't a yes/no thing. It grows with patterns."
A calm reframe:
"My job is to give you freedom you can handle—and help you build the skills to handle more."

5) When You Need Them to Hand Over the Device
This is where many parents accidentally ignite a war.
Try:
"I'm not taking this to punish you. I'm taking it to pause the situation so we can think."
"You're not in trouble for telling me. I just need the device stable for a bit."
If they resist:
"I hear that you hate this. I'm still doing it. We'll talk when we're both calmer."

6) When Something Might Be AI, Fake, or Manipulated
"We don't have to decide if it's real right this second. Let's treat it as unverified and check it."
"What would we need to know to be confident?"
"Is this trying to make you feel something fast—anger, fear, relief? That's a clue."
A line worth repeating in your house:
"Strong feelings are a signal to slow down."

7) When Someone Is Pressuring Them to Keep a Secret
Kids get trapped here because secrecy feels like loyalty.
Say:
"If someone asks you to keep a secret from me that makes you feel weird, that's not loyalty. That's control."
"You will never be in trouble for telling me about a secret someone asked you to keep."

"We can handle hard things together. You don't have to manage it alone."

8) When They're Being Harassed or Targeted
"I'm sorry this is happening. You don't deserve it."
"We're going to handle it, but first we're going to save what we're seeing so it doesn't disappear."
"You're not alone in this. I'm on your side."
If they're ashamed:
"You didn't cause someone else to choose cruelty."

9) When They Did the Harm (Mean Messages, Gossip, Sharing)
This is the moment to be firm without turning your child into "a bad kid."
"I care about you too much to pretend this is okay."
"I'm not saying you're a bully. I am saying this behavior hurt someone."
"Walk me through what you were feeling right before you hit send."
Then:
"We're going to make repair part of this, not just consequences."

10) When You're Not Ready to Talk (Because You're Too Activated)
This is an underrated parenting superpower: pausing without abandoning.
"I'm feeling too charged to do this well right now."
"I'm not ignoring it. I'm taking time so I don't make it worse."
"We'll talk in an hour / after dinner / tomorrow morning. You're safe."
If you're worried they'll delete evidence:
"For now, don't delete anything. Just leave it as-is."

11) When You Need a Second-Channel Verification Habit
Use this when money, pickups, passwords, or urgent requests show up.
"In our family, anything urgent gets verified in a second way."
"If I get a text that seems off, I'll call you. If you get a call that seems off, you'll call me back."
"It's not mistrust. It's how we stay un-scam-able."

12) The Weekly Check-In Opener (Low Stakes, No Interrogation)
"What felt good online this week?"

"What felt gross or stressful?"
"Any app that made your mood worse?"
"Anything you want help fixing or setting up?"
And one that keeps it human:
"What's something you wish I understood better about your online world?"

A Closing Perspective

You don't need perfect words. You need words that keep the relationship intact long enough for the lesson to land.
Most digital harm gets worse in silence.
Most resilience starts with a kid believing: *I can tell my parent, and it won't destroy me.*
That's the quiet win these scripts are built for.

A Final Note, And Where This Continues

If you made it this far, thank you.

That isn't a formality. It matters.

It means you didn't just skim for quick fixes or rules to enforce. You stayed with the harder work: understanding how these systems actually shape attention, behavior, trust, and family dynamics. That kind of care shows up long after a book is finished.

This playbook was designed to be solid on its own. But the reality of the digital world is that it keeps changing. New platforms appear. New risks emerge. New questions surface at moments you didn't plan for.

That's where the website comes in.

It's where the living material lives.

On the site, you'll find:

- expanded versions of the appendix tools
- printable checklists and scripts
- updated guides as platforms and risks evolve
- deeper dives into topics that didn't belong in the main text
- age-specific considerations when you're ready for them
- calm, practical guidance for moments when something feels off and you need a second layer of clarity

Nothing there replaces what you've already read. It extends it.

If this book helped you pause instead of panic, ask a better question, or see your child's digital life with a little more clarity, that work continues there.

No pressure. No urgency.

Just a place to return when you need it.

Scan the code below to continue at SafeScreensWeekly.com

www.ingramcontent.com/pod-product-compliance
Lightning Source LLC
La Vergne TN
LVHW090519110826
845146LV00003B/913